The Evolution of Power

The Evolution of Power

A NEW UNDERSTANDING OF THE HISTORY OF LIFE

GEERAT J. VERMEIJ

PRINCETON UNIVERSITY PRESS
PRINCETON & OXFORD

Published by Princeton University Press
41 William Street, Princeton, New Jersey 08540
99 Banbury Road, Oxford OX2 6JX

press.princeton.edu

Library of Congress Cataloging-in-Publication Data

Names: Vermeij, Geerat J., 1946–author.
Title: The evolution of power : a new understanding of the history of life / Geerat J. Vermeij.
Description: First edition. | Princeton, New Jersey : Princeton University Press, [2023] | Includes bibliographical references and index.
Identifiers: LCCN 2022054721 (print) | LCCN 2022054722 (ebook) | ISBN 9780691250410 (hardback ; acid-free paper) | ISBN 9780691250397 (ebook)
Subjects: LCSH: Evolution (Biology) | BISAC: SCIENCE / Life Sciences / Evolution | SCIENCE / Life Sciences / Ecology
Classification: LCC QH361 .V47 2023 (print) | LCC QH361 (ebook) | DDC 576.8—dc23/eng/20230112
LC record available at https://lccn.loc.gov/2022054721
LC ebook record available at https://lccn.loc.gov/2022054722

British Library Cataloging-in-Publication Data is available

Editorial: Alison Kalett and Hallie Schaeffer
Production Editorial: Karen Carter
Jacket/Cover Design: Heather Hansen
Production: Jacqueline Poirier
Publicity: Kate Farquhar-Thomson and Matthew Taylor
Copyeditor: Karen Verde

Jacket/Cover Credit: Crab, bark, turtle, feathers / Unsplash; Cheetah / Danita Delimont Creative / Alamy Stock Photo

This book has been composed in Arno and Sans

Printed on acid-free paper. ∞

Printed in the United States of America

10 9 8 7 6 5 4 3 2 1

To Edith

CONTENTS

PREFACE

The world in which we live is a world of action, of machines and living things doing work. The work varies and changes, as do the contraptions and organisms doing it; but there is activity always and everywhere in the universe. The curious scientist contemplating this reality is first confronted with the daunting task of describing the seemingly endless variety of agencies that perform work and the consequences of their actions, but the even more challenging step is to establish a coherent framework of explanation, of causes and consequences, relationships and interactions, a common language in which to express and evaluate ideas about how things work and change.

When it comes to machines, physicists and engineers have given us the concept of power, the amount of work done in a given interval of time. Biologists, meanwhile, gave us the concept of evolution—descent with modification—the idea that things (especially living things) change in comprehensible ways according to principles of inheritance and differential survival and propagation. The prospect of joining these two concepts—power and evolution—is therefore especially appealing for what they together might reveal about how organisms have lived and changed over the course of time, and about how the human story fits into this larger picture.

Humans and other life-forms are, of course, not merely machines. Unlike inanimate machines, they work on ways that promote their own interests, notably survival and reproduction. Their power gives them purpose and freedom, qualities imbued by the state of being alive. Power, in short, is what makes life work.

The concept of power as applied outside the realm of machines has long been appropriated by philosophers, historians, sociologists, and politicians concerned with human affairs. Power as conceived in this way might look very different from the concept familiar to engineers, because it is rarely explicitly expressed in terms of energy and time and is instead used metaphorically to describe influence, dominance, and social relations. Nevertheless, power as conceived in the social sciences entails investment in time and work and is therefore usefully interpreted as a socially relevant variant of a widely applicable measure of activity.

Power in the human sphere is wielded and experienced socially. It is about influence and control, status and leadership, wealth and military might, and social hierarchy, all expressible in units of energy and time. Whether it is exercised through coercion and violence or through softer persuasion and magnanimity, power becomes meaningful only in the context of interactions within and among groups. In our species it is projected through speech and written communication and, more overtly, by force. As individuals and groups, we admire and fear power; we wield it, envy it, seek it, submit to it, fight it, and lament its unfairness. Power is a universal reality in the political and economic life of our species and touches every aspect of the private lives of individuals.[1]

Differences in power among sectors of human society lie at the core of conflict, social and economic inequality, trade, and social adaptability. They affect who makes decisions, who has enough resources and time—and therefore power and opportunity—to realize their aspirations, how power can be gained and regulated, and whether and how the status quo of the power structure can change.[2]

With such a human-centric orientation, it might be tempting to think of power as a uniquely human attribute, an inescapable expression of unique human nature. From the ancient Chinese and Greek philosophers to Machiavelli and untold numbers of modern thinkers, power has been employed exclusively in the human context. The implicit assumption animating this frame of mind is that humans stand apart and supreme from all other life; conceptual unity was sought to explain human behavior, but it did not extend to other animals, much less to plants or to microscopic forms of life that were then wholly unknown.

To me as a biologist and evolutionist, however, this human-centric social construction of the concept of power denies our phylogenetic link to the rest of the living world and is therefore far too limited in scope. Scientists like Frans de Waal and many others who have observed behavior in mammals and birds have long recognized that power relations play an important role in the lives of these animals, but the farther we stray from our warm-blooded vertebrate relatives, the less likely people are to think of power as a unifying condition and principle in the realm of all living things. As I shall argue in this book, all life has agency, the capacity—however modest—to use energy and time to compete for resources and to make a future. Power is the property of life that enables organisms to modify or maintain conditions favorable to them. Narrowly defined as work per unit time, power applies to the inanimate and animate realms alike, with the crucial addition that living things apply it to economic ends.

Expansion of the concept of power to all the rest of life is not just an exercise in bringing disparate fields of inquiry closer together. If it is to be useful, the approach should bring to light new insights about both evolution and its productions and about the manifestations of power in human society. Ideally it would clarify whether and how limits to power apply to our species, and how life has coped with and in some cases overcome such limits. The study of nonhuman power, in other words, expands the range of possibilities for the human species to consider as we grapple with the unprecedented monopoly over Earth's resources that modern humans have created.

I shall make the case in this book that the concept of power unifies a vast range of natural phenomena, historical processes and trends, and economic relations in the realm of evolving life. I argue that power has increased over time, giving collective life greater control over its circumstances and opening up previously unimaginable possibilities of structure and change. The study of life's power makes sense of the astounding diversity of forms and functions; it explains the universality of adaptive evolution and the development of antagonistic and collaborative economic interactions and their outcomes. Although power is often associated with coercion and unpleasant consequences, I argue in chapter 9 that it also explains the emergence of beauty, harmony, and freedom, properties that make the world so appealing to those of us willing to observe and understand it. In short, the perspective I develop in this book makes sense of history, not just the seemingly chaotic sequence of events and circumstances of human history but also the saga of life on Earth over the past three and a half billion years. The central role of power in history allows us to see that humans are part of that much larger and longer story. Our actions, spurred by an extraordinary concentration of power, represent an acceleration of processes that have played out for billions of years. The modern human economy is the culmination of long-term ecological trends, not a departure from them. Beyond the fascinating contingent particulars of time, place, and participants, the study of power imbues the history of life with a predictable and discernible trajectory of increasing power.

This may seem pretentious, grandiose, and overly ambitious. After all, most historians, whether they are paleontologists, biographers, or scholars specializing in specific periods of human history, reject grand claims in favor of the details as revealed by fossils, the written record, and phylogenetic evidence inscribed in DNA and in human languages. I, too, revel in these details of who interacted with whom, how individuals and systems of the past worked, the causes and consequences of crises, the opportunities arising from innovations and inventions, and patterns of evolutionary descent. In fact, such particulars are essential to the discernment of broad patterns and the development of an overarching unified theory of life's history. Abstractions achieved by ignoring the properties of organisms and the many ways in which individuals and groups interact and make a living rob history of much of this allure. There is much more to history than dates, names, and numbers. Living entities have characteristics shaped by evolution. They do things; they give history life.

As discussed in chapter 1, many biologists and cosmologists have chosen energy rather than power as the unifying theme in evolution, economics, and the universe. Energy, or work, is equivalent to currency, a means of exchange, but the dimension of time is necessary to make energy do work. Thinkers such as Lee Van Valen, James H. Brown, Eric Chaisson and others who express evolution in terms of energy admit that they often mean power, the rate of energy production or use. Power is thus more inclusive than energy, and is the more appropriate measure of activity.

Why this emphasis on power as a unifying principle of life and its history, and not on other phenomena such as complexity? Unlike power, which enables interaction and implies activity in the realm of living things, complexity is a static condition, in effect a description of structure. In their careful dissection of complexity, Daniel McShea and Robert Brandon distinguish among several conditions that have been lumped under the term complexity. The most common version is the number of, and differentiation among, parts of a whole. The cells of a developing body, for example, differentiate into tissues and organs; lineages arising from a common ancestor diversify and come to differ in their traits, creating not only many lineages but many descendants that vary in their properties. Disparity—a measure of mean difference among the traits of descendant lineages—is thus an evolutionary expression of complexity. Another conception is hierarchy, the number of levels of increasing inclusion in a system of nested parts. In a hierarchy, the most basic parts—those at the least inclusive level of the living system—are often the least complex, having lost their internal differentiation relative to the condition in which these parts were autonomous and not bound to the hierarchical system.[3]

In Brandon and McShea's conception, an increase in evolutionary disparity arises inevitably from a combination of inheritance and the accumulation of errors. Greater disparity over time should therefore apply in almost all circumstances where an ancestor gives rise to descendants, except for extinction and other selective phenomena, which eliminate many lineages and therefore reduce the mean difference in traits among the survivors. A reading of the history of life based on this version of complexity could therefore be a reasonable alternative to the one I favor, an account based on power.

My reasons for rejecting complexity in favor of power as a principle by which to explain life's history rest on three arguments. The first is that, as already noted, complexity in its evolutionary sense is a static property of individuals and of branches in the phylogenetic tree of life. The harnessing and deployment of energy over time—the essence of power—is, on the other hand, fundamental to how organisms make a living. Life cannot exist without power; its activity, structure, and indeed complexity are both determined by, and outcomes of, the interactions mediated by power. To me, therefore, power is more fundamental than disparity or hierarchy as the key to how life works and evolves.

The second argument is that complexity arises not only from the accumulation of errors and subsequent divergence in evolving lineages, but also in ecosystems, in which complexity comes about through the mutual activities of multiple individuals and lineages with many evolutionary origins and adaptive trajectories. Just as organisms must be workable wholes comprising parts that mostly develop by differentiation of a fertilized egg (although there are also many symbionts with different origins in a well-functioning body), an ecosystem is an emergent structure that must function to the benefit of its constituents. An economy's complexity reflects interactions among parts—lineages, individuals, and groups—that change or evolve in response to each other and not autonomously and independently as lineages are assumed to do in Brandon and McShea's conception of how disparity and diversity come about. The constituents of an ecosystem or economy may come together by accident, but the emergent system requires that they accommodate each other in order to remain viable. Such accommodation is active, and therefore not synonymous with self-organization.

Finally, complexity is an abstraction, a condition that by itself does not make the whole alive. Power, on the other hand, introduces the crucial economic dimension, one that affects all living things. Evolution is more than inheritance and descent; it is also modification. Although such change might well begin as error, its fate is dictated by interaction, a fundamentally economic phenomenon whose outcome is affected by the power of interacting participants.

I might also have chosen internal control, or homeostasis, as the key to life and its history. The capacity to buffer a workable body or economic system from outside influences is unquestionably important, as I shall discuss in later chapters; but, like complexity, it is an expression of power, both a consequence and a condition favorable to the evolution of greater power.

Still others have sought unity in evolution by pointing to far-reaching and rare transformations in life's organization and in the mechanisms of change. Innovations such as the cell, chromosomes, sex, photosynthesis, respiration, and a differentiated multicellular body each dramatically increased the adaptive possibilities that life could reach. New symbioses and collaborative arrangements also played their part. Moreover, the early biosphere in which most evolution occurred at the biochemical level was greatly expanded to encompass morphological and behavioral evolution and, with the emergence of humans, technological evolution. Each new level of evolutionary change superimposed entirely new ways of making a living on the preexisting structure. In Enrico Coen's conception, there is a progression from evolution by inheritance to development (differentiation of parts of the body), learning (changing of neural connections through experience), and cultural (or nongenetic) change, where each new phase ushers in what Coen considers an expansion of life's creativity. To me, all of these scenarios and transformations are harmoniously and fundamentally expressible in terms of increasing power. The history of life is an accumulation of innovations that have expanded the capacity to generate and apply power.[4]

Is there a danger that the theme of power is too all-encompassing to remain meaningful as the basis for explanation? In its multidimensionality, power can be expressed in so many ways and be enhanced in so many directions that critics of my approach might conclude that no avenue is left open by which to falsify claims about the history of power. I disagree. As I elaborate in chapter 3, methods exist for testing hypotheses of increase proposed in this book; indeed, these methods identify times and places where power remains low or is reduced. The important step in the scientific study of power is to establish criteria for accepting or rejecting claims about power. The same requirement applies to notions such as evolution, adaptive evolution, the atomic theory of chemistry, particle theory in physics, and the origin of the universe. I see it as the task of synthetic science to build and support as broadly applicable a theory as possible based on verifiable facts and tests (see also chapter 9).

If someone had told me years ago that I would someday write a book about power and history, I would have dismissed the idea as preposterous.

I had no personal desire for power, no aspirations of becoming a politician or a business tycoon, no dreams of wielding power as a university administrator or social activist. Instead, I hoped for academic recognition as a scientist, an evolutionary biologist and paleontologist with a passion for the study of shells and a love of nature. I thought of myself as an observer of the politics and economics of the day, and I wondered about the role of power in human society just as I wondered about the significance of religion without myself being religious. My scientific curiosity ranged very widely, and from the beginning I sought to situate the particulars of my research into a broader context with an approach that might best be characterized as comparative natural history.

Travel stimulated a search for patterns in the geography and history of shell form and function, the architecture of plant leaves, and the shapes and sizes of crab claws. I began to ask questions about the differences and how they came about, what limited the evolution of many plausible modes of life, and how conditions in the distant past differed from those prevailing today. Exposure to new places made me question the familiar as well as the unfamiliar.

Early in my career I realized that predators, by competing intensely with each other for prey, were responsible for the evolution of armor and other defensive attributes of shell-bearing molluscs. Their effects exemplified a process that, borrowing from military language, I called escalation, in which enemies influenced the characteristics of their competitors and victims. This evolutionary process appeared to be episodic rather than continuous, leading to the question how bouts of escalation start and stop. Accordingly, I began to explore not only the evolution of other kinds of plants and animals from the perspective of escalation, but also military history in the hope that the latter might offer clues about the circumstances under which arms races begin and end. The more I read and thought, the more parallels I perceived between human affairs and the rest of the living world. In short, the concept of power and the factors promoting and limiting it emerged as a unifying framework for understanding economics and history.

This book, then, is about what living things do and how effectively they do it. It is a history of performance by individuals and organized groups and of the consequences of power through the ages. Although I shall have things to say about how evolution works, the book is less about the informational underpinnings of evolution or the phylogenetic relationships among the great branches of life than about traits, functions, and the ways organisms make a living and establish sustainable ecosystems. I discuss the properties and consequences of evolutionary breakthroughs but not the genetic or developmental mechanics of innovations. Instead, the book explores the evolving relationships between organisms and environments, with the aim of probing the nature of history itself.[5]

When students come to me saying that they are interested in something—evolution, astronomy, political science, or whatever it may be—my first question is, "What do you want to know?" Readers have a right to ask that question of me. What is it I want to know about power?

From the present perch of humanity in the Anthropocene, it is no exaggeration to claim that no species on Earth, now or in the past, has achieved a level of individual and collective power of *Homo sapiens*.[6] This reality strongly implies that power at all levels of life's hierarchy has increased over time, but it leaves open many questions about how this rise came to be. Was it constant or episodic? Were there downturns? Is it the consequence of fundamental processes and characteristics shared by all life-forms, or is the increase in power mostly the consequence of chance events and circumstances? Is the apparent increase in power observable mainly at the level of individuals, at the level of groups, of ecosystems, or only at the scale of the biosphere, the totality of life on Earth? Or is it discernible at all of these scales? Can power increase without limit on a planet that we now completely control? Would a similar increase in power characterize life on other planets, even if details inevitably vary? Finally, what can the history of power tell us about the choices that lie ahead for our species?

These questions inform the structure and arguments of this book. In chapter 1 I define and characterize power and agency, and suggest that adaptive change and the maintenance of the adapted state are universal economic phenomena resulting from the relentless action of natural selection and activity. The factors that enable or limit the power that organisms can acquire and apply are explored in chapter 2. Limits and opportunities are intimately bound up with the exercise of power through feedbacks between causes and consequences of interactions. With an increase in power, constraints diminish and more avenues for further increases in power present themselves. In chapter 3 I ask how power and its history can be studied, and what statistical and other potential pitfalls might mislead or compromise such an inquiry. Chapters 4 through 7 document the history of various aspects of power: body size (a passive dimension of power), locomotor performance, aggression and other manifestations of force, and rates of production, consumption, and recycling in ecosystems. Each of these chapters explores a different dimension of power. In chapter 8 I consider the place of humans and of the modern human economy in the context of the long history of power in the biosphere, and speculate on the circumstances in which our unique attributes originated and developed. Our economic trajectory, though enormously accelerated and leading to unprecedented global monopoly of the biosphere, represents a continuation of processes reaching back to life's beginnings. The problems and potential solutions of this monopoly take center stage toward the end of the chapter. Finally in chapter 9 I consider the nature of history itself. To what extent is history contingent on particular circumstances, and how did contingency and dependence on the past give way to emergent trends? In this chapter I also discuss freedom and beauty, two qualities we as humans cherish, and how they, like all the other phenomena I chronicle in this book, are the consequence of increasing power.

Before launching into my attempts to answer these questions, I must make clear that I attach no value judgments to power. It is neither all good nor all bad; it just is. Power is an inescapable reality that must be studied and accepted, together with its consequences and limitations. Likewise, I shall have things to say about inequality, including variations in its magnitude and consequences in nature and in the human species, but I am not advocating for or against policies to enhance or diminish it. My goal is to provide the background so that others may use the science to inform policy choices.

Readers familiar with my previous books might legitimately wonder whether I bring anything new to the table in this book. Although I was certainly aware of power and energy as critical dimensions of life, I had failed to grasp the importance of agency, enabled by power, in adaptive evolution. Like most of my colleagues, I accepted natural selection as the primary—some might have said the only—mechanism by which organisms adapt to their environment. While I do not question the essential role that natural selection plays in adaptive evolution, extensive and critical reading in a wide variety of disciplines, together with deep thought and my own experiences in nature, have convinced me that there is more to adaptation and history than selection, and the performance of organisms as they work to acquire and defend resources and forge sustainable economic systems has been missing from most evolutionary discussions. Moreover, important fossil discoveries and interpretations have come to light, and my thoughts on economic issues have deepened and matured, necessitating a reappraisal in some cases. Science sometimes demands that we change our minds, and I have heeded that requirement in this book.

The evolution of my views should not be taken to mean that I embrace a supernatural force or god-like figure as the ultimate cause of the trends I document. To the contrary, I reject such an unknowable prime mover. As observable phenomena, natural selection and agency together suffice to explain the diversity and history of life on Earth, and provide the basis for predicting the nature of life elsewhere in the universe.

Scientific conclusions are by necessity always provisional and incomplete. Not only does new evidence force modifications in our understanding, but many questions remain open and problems remain unresolved. I point to many of these in the chapters to follow. The open-ended nature of the scientific enterprise is one of the enduring attractions of exploring the world and a source of perpetual wonder and curiosity.

This book is a work of empirical science, undergirded by a holistic evolutionary and economic theory, and not a work of philosophy. Concepts such as power, agency, and contingency have philosophical implications, some of which I treat in the coming pages, but I have resisted the danger of delving too deeply into such nebulous concepts as the mind, free will, and related subjects that human-centric philosophers have grappled with for millennia. The definitions I employ are pragmatic as well as descriptive and simple, with the aim of making them applicable to as many living organisms and systems as possible. Despite my impatience with the complex and arcane discourse that characterizes so much of the philosophical literature, I hope that philosophers might gain a perspective on their central concerns that goes well beyond the human sphere.

Finally, a note on presentation in this book. To make the arguments as clear as possible and to avoid cluttering the main text with names, I have relegated citations to the endnotes, with full references given in the bibliography. I have also used the endnotes to highlight controversies and to comment on what I perceive to be misunderstandings as well as my revised interpretations of widely accepted ideas.

ACKNOWLEDGMENTS

This book represents a lifetime of reading, observation, discussion, and thought. No project of this kind is ever conceived or executed in isolation, and this is particularly true in my case. Nearly all of the thirty thousand or more books and papers I have read over the years were read to me while I took copious notes in Braille. My wife Edith, and most recently also my outstanding and gifted assistant Tracy Thomson, have each read and fruitfully discussed with me thousands of these scientific, historical, and even philosophical works. Without them as well as Tracy's predecessors (Bettina Dudley, Janice Cooper, Tova Michlin, and Alyssa Henry), I would have been unable to engage with the scholarly literature given that almost nothing—not even many popular books on these subjects—is produced in Braille or in an audio format. Tracy has also been enormously helpful in tidying up the manuscript of this book and making useful suggestions. It is truly a privilege to have such engaged minds working with me.

The Department of Earth and Planetary Sciences at UC Davis has been wonderful in giving me free reign to pursue my research, writing, and teaching while acknowledging that I am utterly unsuited to administration and committee work. This kind of intellectual freedom is essential and indispensable for someone with my interests and temperament, and I am grateful to UC Davis for allowing, or at least tolerating, my chosen work to proceed with little hindrance.

Edith, my wife of fifty years, has been and continues to be not only one of the most perceptive and analytical people I know, but also an unfailing companion and fellow observer in the field and in museums. We share a love of music, a reverence for nature, the pleasure of fine walks, good meals and wines, and an accomplished daughter Hermine and her family. It is with the deepest love and admiration that I dedicate this book to her.

The Evolution of Power

1

The Nature of Life and Power

The central thesis of this book is that power is a universal attribute of life and that the history of life on Earth can be meaningfully and informatively interpreted as a history of power. This claim rests on two concepts—life and power—that are represented by common everyday words but whose precise meaning deserves careful scrutiny. It is therefore essential to specify how life differs from non-life, to explore the unique properties of living entities, and to define and characterize power in the context of life.

Most of us can recognize life when we encounter it, at least when it is in the form of plants and animals. Beyond these familiar versions, however, there is a vast diversity of microscopic life, to say nothing of extinct organisms. Individual living things range over twenty-one orders of magnitude in body mass and represent millions of distinct lineages, each the product of an unbroken line of descent from the origin of life 3.8 billion years ago to the present. Despite this almost incomprehensible variety, all life shares properties that distinguish it from non-life.

To encompass the diversity of life-forms, we need a definition of life that is descriptive, incorporating its unique properties while avoiding inferences about the processes and mechanisms that explain how life works. At its most basic, life comprises particles of matter that convert free energy, reactive carbon, and other substances into structure and activity, properties that are self-sustaining thanks to chemical cycles of renewal and to the potential for the particles to replicate using information encoded in the molecular sequences of polymers. Life is an emergent state of matter because its particles possess characteristics that their non-living components do not possess. As Stuart Kauffman notes, life obeys and is consistent with the laws of physics, but it obeys additional laws that are not reducible to them.

One of those emergent properties of living things is agency. By agency I mean activity, or doing something. Agency entails expending energy and time, that is, power. To many philosophers, the concept of agency implies intentionality, or awareness of action; but this is a narrowly human-centric view of a property that characterizes all life. Agents—living things that have agency—do things with or without conscious knowledge of what they are doing. The key to the idea of agency is that organisms affect themselves and their surroundings. In J. Scott Turner's apt words, life "makes the future happen." By converting energy and matter into activity, living entitles interact with each other and affect their own and others' environment, including resources. Bacteria can climb gradients of chemical concentration. Animals move toward more favorable conditions or away from danger. Plants and fungi grow in directions that afford greater access to resources, and vines' tendrils and growing tips actively move in search of supports. Many organisms expend energy to regulate and stabilize conditions inside the body, affording them a degree of autonomy from their surroundings. For many animals, including humans, agency can become intentional, involving deliberate action with a perceived goal and purpose. There is in fact a fine line between "unconscious" activity and purposeful intention.[1]

The concept of agency as I use it in this book has a long history dating back to Aristotle, if not further. Aristotle employed what he called *psuche* to name what we might label vitality, or the force of life, an attribute that inanimate matter lacks. Although he could not characterize *psuche* and made no predictions about his concept, it was obvious to him that living beings do things, an activity that in modern terms requires the application of power.[2]

Activity that makes the future happen requires that an organism perceive and respond to its surroundings. This engagement with the environment in turn is impossible without power. Most forms of sensation—vision and light perception, hearing and related perception of mechanical vibrations and other stimuli, chemical detection, and electrical sensation—depend on either the sender or the receiver, or both, to move or to generate a signal. Animals moving through water, for example, create disturbances in the surrounding medium that are detectable by other animals.[3]

Response likewise entails the deployment of power, often in the form of movement of part or all of an organism's body. The compound leaves of members of the pea family (Fabaceae), for example, change from a horizontal orientation during the day to a more vertical one at night thanks to the metabolic activity of pulvini, structures at the base of each leaflet. The leaves can fold within minutes of being mechanically touched, again because of power generated in the pulvini.[4] In many land plants, opening and closing of the stomates by guard cells on the leaf surface regulates gas exchange and transpiration, and is a metabolically active process that differs from the more passive opening and closing of stomates in other land plants. The understory herb *Elephantopus elatum* (Asteraceae) growing beneath pines in the Florida savanna has broad, ground-hugging leaves that use forces of up to 0.058 newtons to push grasses and other competitors away. Plants with the habit of producing a basal rosette of leaves are common, raising the possibility that such behavior against competitors is widespread.[5]

Agency also characterizes developments within organisms. Motile cilia orchestrate cell movements within the developing body of most organisms, and the twisting and curving that are the hallmarks of early ontogeny in many animals, reflect the action of mechanical forces. It is far beyond the scope of this book to explore these dynamic forces. Suffice it to say that curves and deformations of growing body parts as the parts interact with one another are universal features of organisms.

Whichever form it takes, agency provides the potential for a living thing to survive long enough to leave offspring. In other words, it is the means to achieve success, as indicated by survival and reproduction. Persistence and perpetuation, in turn, depend on performance, that is, on how well life translates resources into agency and agency into resources.

The phrase "how well" implies a comparison, in particular a comparison with the performance of other living things.

Put another way, performance is a measure of, or at least reflects, competitiveness. Polymers compete for their monomeric components in a prebiotic world; parts of the brain compete for space in the head; and organisms compete for all sorts of resources. Whenever there is more than one living particle or part of one tapping a resource, there is competition for that resource on a local scale. Trees compete with other plants for light, water, and nutrients, as well as with insects and mammals that eat their leaves. Lions vie with each other for prey and with hyenas that steal hard-won quarry, and with the parasites in their digestive systems that tap some of the nutrients acquired by the lions' efforts. Swallows compete for holes in cliffs or riverbanks, so that they can build nests where their eggs can develop and hatch in relative safety.

Animals that practice internal fertilization, in which one individual mates with another internally or in a controlled space where eggs and sperm meet outside the body, compete fiercely with, and choose among, potential partners. Flowering plants compete for pollinators, which act as surrogate vectors for sperm, by attracting them with nectar and other benefits. In its many forms, including working with others cooperatively and mutualistically as larger units, competition for locally scarce resources is a fundamental activity of living things.

The structure of organisms is to a large extent specified by their genomes, comprising molecular sequences (genes) of DNA and RNA that encode and regulate the formation of proteins, which in turn are long chains of amino acids. Whereas agency makes the future happen, the genome is a record of past success. Replication of genes inevitably introduces errors, which create genetic variation that in turn affects performance. As living things compete for locally scarce resources, some are better at it than others. They acquire more of the necessities of life, defend resources and their own bodies more effectively, and convert energy and matter into structure and activity so that survival and reproduction are more likely. Competition implies agency, but in self-replicating organisms its outcome is also affected by genes. For this reason, competition is the economic foundation of natural selection, the genes-based process that, through differential survival and reproduction, maintains or improves adaptation, the good fit between a living thing and its environment. Natural selection operates at the level of genes, but it is the whole organism that has agency, and it is the living whole that does or does not survive and reproduce. Natural selection therefore acts on the whole agent. Together with activity, natural selection is the basis of adaptive evolution—descent with adaptive modification—which is yet another fundamental property of living organisms. The good fit between an organism and its circumstances is thus the consequence of two distinct processes, natural selection and agency.[6]

It is important to make a distinction between evolution in general—descent with modification—and adaptive evolution. The former is a property of all systems, living as well as non-living. These systems change over time but not necessarily in adaptive ways and not necessarily involving replication. Galaxies, stars, climates, Earth's crust and atmosphere, and the universe as a whole can meaningfully be said to evolve, even in directions that are predictable, but they do not maintain or improve a good fit with, or accommodate, their environment. Adaptive evolution is therefore a property unique to life, arising from natural selection among replicators and from life's agency.[7]

To see how both agency and natural selection are necessary for adaptation, consider the case of camouflage, the close visual resemblance between an animal and its surroundings. The trait that makes the animal hard to see is controlled by genes, but if the animal found itself in an environment where the camouflage does not work, it will be quickly found and consumed by a predator. The animal must therefore use its agency to locate and remain in an environment where it will be least conspicuous to its enemies. This requires sensory capacity as well as mobility on the part of the animal. Although these traits are certainly also influenced by genes, their realization requires a combination of stimulus and suitable response or agency.

Adaptation is rarely if ever perfect. It persists because it works well enough under the circumstances, not because it is optimal. A visually cryptic moth on a tree trunk need not perfectly match its background, but it must fool a potential predator long enough to avoid being eaten.

An orchid flower need not precisely mimic a female wasp, but the resemblance must be good enough to entice a male wasp to visit, attempt to mate, and thereby pollinate the flower. There is thus always room for improvement, depending on how intense competition and natural selection are.

The more agency an organism has—that is, the more power it wields as it interacts with other life-forms—the greater is its capacity to influence and respond to its surroundings. Performance is power, and adaptive evolution is the long-term outcome of short-term interactions between living entities.

Watt Is Power?

The concept of power as used in this book has a specific meaning, one borrowed from physics and engineering: Power is energy per unit time, expressed in watts. Energy, or the molecule harnessing it in organisms, is a currency, or means of exchange. Raw materials necessary for the construction of living things, together with free energy, must be both available and accessible to organisms. Like money in human society, energy by itself has no value to a living entity unless it is spent, that is, unless it performs work. Without energy, nothing happens and nothing gets done. The value of energy—and money—lies in its being put to use, and that entails time and therefore power.

Power can be wielded not just by cells and individual organisms, but also by organized groups, ranging from populations and communities to ecosystems and the global biosphere. The potential to grow—still another trait common to living things though not unique to them—is one way, but not the only way, in which living entities at all scales of inclusion can increase their power. Whether that potential is realized, and whether power in general exhibits trends over time, depends on external circumstances, internal technology and innovation, and the feedbacks between what economists call supply and demand, or what an evolutionary biologist might call opportunity and constraint. I deal with these issues in chapter 2. How power and the factors controlling it change over time is the central question examined in this book.

Table 1.1. Equations for power and its components

Power: energy per unit time (watts)
$P = mv^2t^{-1} = md^2t^{-3}$
Energy: force times distance (joules)
$E = mv^2 = mad$
Force: mass times acceleration
$F = ma$

Symbols
a = acceleration, in meters per second squared
d = distance, in meters
m = mass, in kilograms
t = time, in seconds
v = velocity, in meters per second

The dimensions of power offer a guide to the many ways in which organisms can increase their power or diminish the power of others. The equations in table 1.1 express power in distinct but equivalent ways, each emphasizing a different variable that competition and natural selection can target. There are thus many pathways toward greater power and toward a denial of power, and living things have explored them all.

From the perspective of a living entity, then, power might increase because of greater mass (m), greater force (F), higher velocity (v), faster acceleration (a), greater distance traveled (d), larger area covered (A), and a shorter time (t) in which a given reaction or activity takes place. Moreover, living biomass can be divided among multiple entities; therefore, power can increase as the number of entities increases in a fixed space. Put another way, abundance of living things is a component of power at the group level.

Of all the dimensions of power, time is most important, because it appears with an exponent of negative 3 in equation 1 in the table, whereas other variables have exponents of 1 or 2. A change in time therefore has a greater effect on power than change in variables such as mass, area, distance, and velocity, in which time either does not appear at all (mass, area, and distance) or has an exponent of negative 1 or negative 2. In short, the dimensionality of power tells us which adaptive possibilities for increasing or thwarting power are available to competing life-forms.

An interesting example of how a plant manipulates the time and power of its insect adversaries comes from hairy leaves and stems. I had known for years that the hairs of many plants are oriented in such a way that they point toward the tip, or apex, of a leaf. In the same plant, hairs on the stem often point toward the ground. It occurred to me that one possible function of asymmetrical leaf hairs might be to direct the movements of a caterpillar or other insect toward the leaf's tip, so that the insect spends less time chewing the leaf as the plant shepherds the intruder off. Downward-pointing stem hairs, by contrast, discourage climbing insects from ascending the plant, increasing the time it would take a caterpillar or aphid to reach the palatable parts of the plant. In experiments with oats (*Avena barbata*) and the small caterpillar *Heliothus virescens*, we found that the insect crawled preferentially toward the leaf tip as expected from the apex-pointing hairs; but the larger caterpillar *Arctia virginalis* did not show this behavior.[8]

Different names are assigned to power according to which aspects of power are being measured and which kind of living entity is involved. For individual organisms, metabolic rate (the amount of energy acquired to deployed per unit time) or fitness (the number of offspring produced per unit time or over a lifetime) are suitable measures of power. For populations, ecosystems, or human-economic units, power might be expressed as productivity, the rate at which biomass (or economic activity) is produced per unit time. Other variations might include the rate of growth, the rate of predation, the rate at which the environment is being modified, and influence, the effect that one entity has on the activity of another. The important point is that all these measures have in common their expression in units of power. Performance is power; and energy is currency.

Life is a perpetual struggle against disruption and chaos. In its many manifestations, life can be thought of as an ordered concentration of power, a local and temporary violation of the increasing entropy that physicists assure us characterizes the universe as we know it.[9] Life flies in the face of static equilibrium; the greater its power, the farther from stability life strays. Moreover, as its collective power expands, the scale at which life concentrates power also increases, spreading its influence and, as I shall argue, diminishing the destructive effects of external disruptions.

The property of self-perpetuation sets life apart from non-life. The biosphere has persisted for at least 3.8 billion years despite great crises as well as profound environmental changes of life's own making. This is possible only if resource supply is regenerated. This regeneration involves cycles of production and consumption, implying a degree of cooperation among competing entities. The result is what we call an economy, or ecosystem, comprising entities that compete and collaborate, establishing relationships that stabilize and potentially increase the flow of material necessities. The power of global life derives from the feedbacks between resources and living entities and on the capacity to renew and replenish what is used up.

Where Do Agency and Power Come From?

Ultimately, all of life's agency derives from energy and raw materials from inorganic sources. In the simplest case, energy is taken up passively, and work is accomplished with minimal effort by the organism. The ascent of water from the roots through the stem to the leaves of plants, for example, is due to transpiration, the loss of water vapor through pores (stomates) in the leaves. This is an inevitable consequence of photosynthesis.[10] Passive filter-feeders such as sea lilies (crinoid echinoderms) extend their food-collecting structures so that they intercept food-laden water currents.[11] Evaporation and the resulting air flow beneath the cap of a mushroom are responsible for catapulting spores, which are on average 10 micrometers in diameter, away from the mushroom.[12] Seeds with extended hairs, feathery plumes, or stiff wings disperse passively in the wind. Many plants rely on the wind to disperse spores or pollen.[13] One of the few examples of wind as the motive agency for animals comes from the pelagic hydrozoan *Velella,* which floats and moves over the ocean surface with the aid of a sail.[14] Most seaweeds, which must acquire nutrients passively from the surrounding water, grow more quickly in wave-swept environments, because they intercept more water per unit time than they would under calm conditions.[15]

A particularly intriguing example of harnessing the wind to adaptive ends is furnished by poplars and aspens, trees of the genus *Populus*. Their vertically held leaves tremble in the slightest breeze, collectively making a very characteristic hissing sound that can be heard at a distance of tens of meters from the tree. As the leaves flutter in the wind, their smooth surfaces cause small insects to lose their grip and fall to the ground.[16] I have wondered whether a similar antiherbivore effect might explain the clattering sound that the fronds of coconut palms (*Cocos nucifera*) make in the wind. In Dutch, the coconut palm is known as the klapperboom, an apt description. One problem with using power from wind or water currents is that it cannot easily be stored. For effective use, it must be reliable, reasonably consistent, and not so intense that organisms are swept away in uncontrolled ways.

These examples make clear that, even when the acquisition of resources is passive, organisms must nonetheless put energy into the appropriate structure, molecular machinery, and physiology to take advantage of those resources. This investment requires metabolism, a universal capacity of living things that entails the ability to harness external sources, store them in molecules as chemical energy (mostly chlorophyll and adenosine triphosphate, or ATP), and release them in a controlled way. The rate of metabolism—a measure of power—affects how much force, work, and power an organism can produce. The energy content of an organism is that organism's energy budget; the rate of metabolism, or perhaps more accurately the metabolic scope, is the organism's power budget. By scope I mean the difference between the basal metabolic rate, the rate of energy use when the organism is at rest and not engaged in strenuous or expensive activity, and the maximum rate, which can be achieved for short periods of intense activity. For this reason and for others, an individual's metabolic rate is not constant; it varies according to life stage, time of day, temperature, and frequency and intensity of costly activity. Most organisms exercise substantial internal control over metabolism, just as they do over such other functions as growth, locomotion, digestion, respiration, and reproduction.[17]

Movement of part or all of the body is a widespread way in which organisms exert power. It involves motors: the molecules that drive stiff flagella in bacteria, cilia in animals and single-celled protistans, contractile muscles in animals, and the pulvini of plants. Scientists working at the intersection of biology and physics have discovered surprising regularities in the scaling of power and force of motors with mass, either of the motor itself or of the body it powers. Maximum speed of locomotion, for example, scales with body mass to the 1/6 power, regardless of whether the moving structure is an animal or a human-made vehicle weighing less than 35,000 kilograms. Organisms ranging over nineteen orders of magnitude in body mass have maximum speeds averaging ten body lengths per second, all else being equal.[18]

These scaling relationships reflect central tendencies, but they obscure large variations. To a physicist, these variations might simply reflect unwanted noise, but to an evolutionary biologist they indicate that organisms have considerable adaptive scope. In other words, the regularities are not so strict that they constrain the adaptive options available to organisms faced with particular challenges. The variations in actual force and power therefore show how basic rules can be bent or even broken in the course of adaptive evolution.

An example will illustrate the importance of variation. As I noted earlier, organisms ranging from bacteria to large vertebrates have average maximum speeds of ten body lengths per second, measured at activity temperatures typical for these organisms. Some heat-loving archaea—microorganisms that have evolved separately from bacteria at the prokaryote level of organization—have maximum speeds of 400–500 body lengths per second, propelled by flagellum-like structures called archaella.[19] These minute specks of life are thus the fastest known organisms, at least by this metric. Absolute speeds are slow, but in a world where movement of tiny organisms is dominated by viscous forces, they are remarkable.

As generators of power, motors in highly mobile animals can take up a large fraction of body mass. In tropical butterflies that are highly palatable to birds, for example, there is a premium on rapid, erratic flight to avoid being caught by the faster birds. These butterflies devote almost half of their body mass to flight muscles.[20] Pacific salmon of the genus *Oncorhynchus* weighing 0.22 kilograms and capable of burst speeds of six meters per second have 0.1 kilograms of swimming muscle mass, or almost 60% of body mass.[21] These percentages do not include the motors responsible for supplying the muscles with oxygen—in the cases discussed, the heart—or the metabolic costs associated with the brain, which must coordinate sensory and motor functions.

Energy storage in organs dedicated to that function is key to the projection of power in many organisms. Often this is in the form of chemical energy, which can be tapped during times when resources or external energy are scarce. Plant bulbs, tubers, above-ground succulent stems, and seeds rich in oil or starch are familiar examples. Fat reserves enable bears and ground squirrels to hibernate during long winters at Arctic latitudes and enable birds to migrate between wintering and breeding grounds. Bird chicks have access to abundant yolk and albumen inside the egg before they hatch.

Energy can also be built up slowly and stored as potential energy in mechanical devices such as springs before it is released in enormous bursts of power by tripping a latch. Such mechanisms are thus power amplifiers by which organisms, as well as many human-made devices, can generate huge forces, velocities, and accelerations over extremely brief time intervals.[22]

Spectacular examples come from plants that hurl their seeds away from the parent as the fruit holding the seeds fractures explosively. In the herb *Cardamine hirsuta*, a member of the cabbage family (Brassicaceae), the two-valved fruit (technically a silicle) springs open as the valves separate, releasing the seeds at speeds of more than ten meters per second over an interval of three milliseconds, enabling the seeds to travel a distance of about two meters. The stored energy that makes this release possible is due to metabolically generated turgor pressure in the valves, aided by various modifications at the cellular and structural levels in the valve walls.[23] In Panama, the tree *Hura crepitans* (Euphorbiaceae) uses similar mechanisms of explosive fracture to propel its seeds at a speed of seventy meters per second when pressures in the pod exceed ten atmospheres.[24] The spitting cucumber *Ecballium elaterium*

(Cucurbitaceae) and dwarf mistletoe *Arceuthobium* sp. (Loranthaceae) pressurize their fruits osmotically. At a critical pressure, the fruit releases a jet of liquid and mucilage-covered seeds, aided in the case of the mistletoe by metabolic heating of the fruit.[25]

Many small animals also apply amplified power to execute lightning-fast motion. The planthopper *Engela minuta* (Dictyophoridae) weighs about twenty milligrams and is less than one centimeter long, but its long hind limbs release a spring that allows the insect to jump with a velocity of 5.5 meters per second within just 1.2 millisecond.[26] Using a hammer-like finger on one of its mouth appendages, the tropical mantis shrimp (stomatopod) *Odontodactylus scyllarus* smashes the shells of prey snails with a force of up to fifteen hundred newtons at a speed as high as twenty-three meters per second once enough energy has accumulated for release by the fifteen-centimeter-long animal.[27]

A species of *Dulichiella,* a tiny amphipod crustacean found among seaweeds in North Carolina, sports an enormous claw in the male, averaging 30% of the animal's mass. This claw releases a latch so that it can close the one-millimeter-long, 184-micrometer-thick, movable finger of its claw to reach a velocity of 1.7 meters per second in as little as fifty microseconds. Although the function of this extraordinary snap is unknown, the fact that males but not females have the enlarged claw indicates a role in sexual selection, although defense also remains a possibility.[28]

At the high end of behavioral sophistication, some animals resort to storing resources outside the body. Jays, squirrels, and many smaller rodents cache seeds and nuts, and some desert ants store vast quantities of seeds underground. Bees make and store honey. A truly extreme form of external storage is agriculture, in which animals grow fungi or, in the case of humans, plant crops. I return to these remarkable enhancements later in the book.

Why Be More Powerful?

The advantages of wielding more power are obvious and universal. In the short term, powerful agents have greater access to resources and are in a better position to survive under circumstances favorable to them, in part because they are not as threatened and constrained by rivals as are weaker agents. Dangerous and powerful predators, from octopuses and sharks to eagles and wolves, can freely roam uncluttered environments where prey are most abundant and vulnerable. Less powerful competitors are relegated to times and places where resources are less plentiful or accessible. Even for these weaker agents, however, there is an advantage to being as powerful as circumstances, including their superior competitors, allow.[29]

In competition there are always winners and losers in the sense that one party gains more or loses less power than the other. Despite this universal inequality, both parties incur costs. Predation, for example, in which one of the competitors eats part or all of another organism, is potentially expensive for the predator, which must expend power to locate, catch, and subdue its prey. The prey often survives such attempts, owing in part to its costly antipredatory adaptations. In our laboratory trials with the shell-breaking, sand-burrowing crab *Calappa hepatica* in Guam, for example, we found that more than 90% of attacks on well-armored marine snails were unsuccessful, allowing the prey to survive and to repair the shell damage inflicted by the powerful pincers of the crab.[30] All the prey species have specific crab-resistant shell features such as a narrow opening, crack-stopping ribs, and a high spire so that the head and foot can withdraw so deeply into the shell that the predator cannot reach them.

A broad survey of predators ranging from sea stars and planktonic copepods to spiders, snakes, and lions revealed that attackers almost never had a 100% success rate under field conditions. Often these predators would be effective at detecting prey, or catching mobile animals, or subduing dangerous or well-armored prey, but never in all phases of an attack. The cost of failure for predators can therefore be high, although on average not as high as for the prey that are successfully killed.[31]

Another dramatic example of the high cost of being a strong competitor comes from trees. In order to compete for light, trees must grow into the canopy, where sunlight is most plentiful. Given that a canopy already exists and that young trees typically begin life near the ground, the great height necessary to reach the canopy requires wood. In tropical forests, each hectare supports about three hundred tons of plant biomass, of which about 97.3% is wood.[32] Although this is an average based on hundreds of species, including lianas with less wood, the extraordinary allocation of trees' resources to the production of inert wood illustrates the very high cost of being a vigorous competitor in one of the world's most productive ecosystems. Understory plants that are much less productive cannot hope to match such investments.

The inequality of power between competitors has profound implications for adaptive evolution. Given that winners are more powerful than losers, there is a fundamental imbalance in agency and natural selection between the winning and losing sides. Winners have a greater evolutionary effect on losers than losers do on the winners. For predators, this means that competitors for prey exert greater influence on the characteristics and distribution of these predators than do the prey. The sharp claws of lions, for example, are effective weapons for capturing and subduing prey, but selection for sharper claws may be due to competition with hyenas, vultures, and other lions rather than to the defensive strategies of the lion's mammalian prey.[33] This asymmetry of power can be reduced but not eliminated by prey that pose a real danger to their potential predators, as occurs when the prey are venomous, toxic, or aggressive.[34]

This effect, where the more powerful agents have greater influence on the characteristics and distribution of organisms than the less powerful, results in the evolutionary process I call escalation. The idea is that enemies have greater agency and exert more intense selection than victims. It is this enemy-directed adaptive evolution that is responsible for the relentless and pervasive selection for as much power as individuals and populations can sustain.[35]

An instructive example of costs associated with competition and escalation comes from adaptive tail loss in geckos. Like many other lizards, the geckos *Mediodactylus kotschyi* and *Hemidactylus turcicus* from islands and mainland habitats in the Mediterranean region shed (or autotomize) their tails in response to danger. Autotomy enables lizards to escape despite the costs of reduced locomotion that the loss and subsequent regeneration of the tail incurs. Predatory vipers are important agents favoring autotomy, but on islands the primary culprits are members of the same gecko species.[36] Both predators and competitors of the geckos thus elicit expensive adaptations even when the target geckos are not competitive winners.

As long as there are safe places for potential victims, or limits to power of the victors, the outcomes of competition allow both parties to maintain or increase their power through adaptive evolution. There is, however, one important exception to this generality, a case where one side has evolved so much power that the victims' adaptive options disappear. I finally became aware of this possibility after the archaeologist Curtis Marean approached me with a question: why do South African shell middens, which represent the remains of early human marine shellfish diets, show a trend toward smaller shell sizes from lower to higher levels in the middens? It occurred to me that humans are unusual among animals in targeting large prey individuals of species that we hunt, fish, and gather. Reams of data show that people select larger prey over smaller ones.[37] The consequence of this deliberate selection is that the protective and competitive benefits that come with large size are effectively eliminated, leaving almost no scope for adaptation of the prey in response to us as predators. The only remaining option, which has in fact been widely adopted by heavily exploited fish species, is to mature as smaller adult sizes and to reproduce at an earlier age. This trend leads to lower per-capita fecundity, a problem exacerbated by laws that prescribe minimum sizes for harvesting. The same principle applies to the limpets, mussels, turban snails, and other shellfish in those South African middens, and to the harvesting of shore animals worldwide.[38]

This human example exposes an important point about the limits of power. In part thanks to technology and our social nature, we humans have wielded more individual and collective power than any other living or extinct species. Other top competitors have by comparison exerted limited power. This constraint, imposed by diffuse regulation from many species, permits all players to adapt regardless of their competitive status.

I shall leave the question about how humans were able to achieve such unprecedented power until later in the book, but one widespread agency that might have contributed at least to the early ascent of power in humans as well as in many other animals is mate choice. Sexual selection associated with mate choice has led to all manner of extravagant behaviors and bizarre traits, which serve to advertise to, coerce, attract, and control potential mates.[39] I suggest that selection for greater power is amplified in species in which competition for mates is important. The antics and displays are very costly, as the example of the fast-closing and very powerful claw of the small amphipod discussed earlier demonstrates. Such extreme investment indicates that individuals engaging in mate competition have power to spare and, insofar as it confers an advantage in reproduction, it indicates intense selection for greater power. The development of potent weapons like horns, antlers, and claws in animals or large showy flowers in plants testifies to the generality of these advantages in organisms engaged in mate competition.[40]

The hypothesis that mating-related displays indicate overall vigor and "good genes" has long been popular among scientists who study animal behavior.[41] Richard Prum disputes the idea that ornaments related to sexual selection are "honest" signals of overall fitness, presumably because, in his view, survival and fecundity would be higher in their absence.[42] This criticism, however, is beside the point. If displays result in mating, that is what matters, regardless of expense and what might have been. In short, the behaviors and ornaments work. They amplify activity and selection for individual power. In fact, I would turn the "good-genes" hypothesis on its head: It is mate selection and mate competition that indirectly also favor traits that expand power in other aspects of an individual's life.

This argument applies with equal force to plants. In order to attract mates, flowering plants must invest heavily in nectar and other rewards for obtaining the services of faithful pollinators. Such requirements might make selection for greater power imposed by herbivores and by neighboring plant competitors more intense. In this connection it is no accident that high rates of photosynthesis, elaborate mobile chemical defenses such as latex, and the active opening and closing of stomates and of leaves in response to environmental conditions characterize flowering plants but not other plant groups. The idea is that mate competition—in this case for pollinators—favors the increased scope for power that higher rates of production and greater investment in defense provide.

In short, selection for greater power is extremely widespread. Great power may be reserved for the tiny minority of top competitors, but the benefits of power accrue to any organism that has agency. Lineages of losers may initially decrease in power as they become specialized to live in places where organisms are relatively safe from powerful rivals and where resource supply is limited. Once they have adapted to these low-powered modes of life, however, organisms like parasites, cave-dwellers, deep-sea animals, and understory forest herbs must wield as much power as circumstances warrant because they must confront enemies even under these safer surroundings. Indeed, as I shall show in later chapters, innovations in these initially safe environments can enable some lineages to gain enough power that they make these habitats more competitive. Selection for greater power is relentless and universal. It may be more intense for some lineages than for others, and most lineages never achieve the status of top competitor, but power is the ultimate evolutionary aphrodisiac.

2

What Limits Power?

If competition is the economic process that fuels the relentless pursuit of power, it also implies local scarcity and limitation. Both the availability of resources and access to them are subject to global constraints imposed by Earth's chemical composition and energy environment, and to global as well as local limits imposed by life itself. These limitations are not absolute. Through physiological and structural innovations and the establishment of cooperative arrangements among life-forms, many evolving lineages have overcome or at least lessened constraints on power. To understand the evolution of power, it is therefore necessary to probe the nature of constraint on it and to identify the ways in which life has created opportunity by reducing limits on power.

Organisms need energy and raw materials. Not only must these be available, but exploitation also requires access. Availability is dictated by the organism's environment; access is also determined by external circumstances, but it is to a large extent determined by the organism's own equipment and power. In this chapter, I consider how availability and access limit and enable the acquisition of power. In particular, I discuss what I call enabling factors, that is, resources and the conditions affecting their availability and access. Enabling factors control supply, whereas competition in its broadest sense constitutes demand. Supply and demand are linked through positive as well as negative feedbacks. Sometimes they oppose each other and reinforce limits on power; often, however, positive feedbacks ensure that supply and demand can both rise. I argue that, because resources are often alive themselves, controls on them are also exercised by organisms. This simple fact has the profound implication that supply and demand are both subject to adaptive evolution. Together they control an evolving biosphere and create an economic history of life.

Resources

It takes resources and access to them for living things to acquire power, and it takes power to exploit resources. These two intersecting requirements reveal a fundamental truth: Life and environment affect each other through feedbacks. The environment provides opportunities and imposes constraints; competitive processes and activity are responsible for the projection of power and underpin adaptive genetic and cultural evolution. How much power living things can muster depends on how opportunities and limitations affect competition and vice versa. Causes and effects intertwine, and all the moving pieces of the life-environment puzzle are subject to change.[1]

At first glance, there might appear to be only a few basic resources for which organisms compete. Light, water, and nutrients—especially nitrogen and phosphorus—come to mind for land plants; the uptake of carbon from water is a limiting step for many aquatic plants. Food is an essential resource for animals and fungi. Animals that build skeletons of calcium carbonate ($CaCO_3$), calcium phosphate (or apatite, $CaPO_4$), or silica (SiO_4) are often limited by the supply of dissolved components of these minerals. Beyond such essentials, however, there is a vast diversity of resources and services for which organisms compete. Animals and plants compete with their consumers and thus have evolved a spectacular array of passive and active defenses to protect their bodies. Animals and plants that practice internal fertilization engage in intense mate competition, where one or both parties choose among potential mates. Corals, barnacles, and human societies compete for space, and humans compete for status while businesses compete for customers. The diversity of targets for competition therefore enables, and is enabled by, a vast number of ways of making a living. Life is wonderfully multidimensional.

On the global scale, resources such as sunlight, space, and nitrogen are so abundant that ecologists often claim that organisms do not compete for them. Land plants, for example, use only 0.7% of incoming solar radiation, which on the present-day Earth amounts to an average of about 340 watts per square meter of surface.[2] Global availability, however, is a poor indicator of local competition. Organisms, after all, compete locally under conditions where an abundant resource can be scarce. In most forests, for example, only 1% of the sunlight arriving in the canopy reaches the forest floor, because layers of leaves intercept direct sunlight before heavily shaded understory plants can gain access to it.[3] On coral reefs, there is plenty of space for attached organisms to both settle and grow. Space on that scale might limit the number of individuals within a population but not the size of the individuals themselves. Competition between individuals or between colonies, however, is local. Two neighboring coral colonies often fight fiercely over the space between them by deploying long tentacles, sometimes resulting in a no-coral zone between them or enabling one colony to overgrow the other.[4] Similar contrasts between global abundance and local scarcity apply to oxygen for animals—think of life in deep burrows, where oxygen can be in short supply—and to nitrogen for plants, water for trees, dissolved iron for oceanic plankton, and books for readers in the library.

Nevertheless, global supply matters. Consider, for example, the suspended food particles on which filter-feeding animals in water rely. Lineages in almost every major group of animals have become specialized to filter particles from the surrounding water. These range in size from tiny copepod crustaceans to sponges, clams, slipper limpets, barnacles, basking sharks, and mysticete (baleen) whales. Where planktonic food is abundant, clams and barnacles can reach much larger sizes than members of the same species in nutrient-poor waters. Although competition between filter-feeders is local, the abundance of resources on a large regional scale sets limits on how much resource a competitor can expect to acquire over its lifetime. In other words, global availability becomes important on the timescales of lifetimes and beyond, because the results of individual episodes of local competition are cumulative.

Productivity

Ultimately, living things must acquire energy and raw materials from non-living sources. The sun and the Earth's internal heat provide energy; the atmosphere, oceans, and crust are the sources of carbon, oxygen, nitrogen, sulfur, phosphorus and all the other elements of which organisms are made. At least in the early phases of life's history, therefore, the availability of the ingredients of living matter and activity is determined by celestial and geological processes and conditions. Although these controls remain important in today's biosphere, the effects of life itself on resources have modified and often stimulated supply to the point where the relationship between life and environment has achieved a measure of autonomy from the initial external controls. In other words, life has collectively achieved power over itself.

This fact has profound implications for life's persistence on Earth. Living organisms and the organic matter they produce and consume form most of the food base and chemical energy for living things. Just as organisms must on average replace themselves with their offspring, the nutritional foundation must also be collectively replenished by organisms. Plants, seaweeds, and phytoplankton are the primary producers, which fix carbon from carbon dioxide or the bicarbonate ion (HCO_3^-). Animals are the primary consumers of plants as well as the secondary consumers of other animals. Fungi and microorganisms, sometimes aided by scavenging animals, transform dead organic matter into minerals that the primary producers need. An overall indication of how rapidly this cycle runs—that is, of the ecosystem's power—is net primary productivity, the rate (power) at which plants fix organic carbon that is consumable by other organisms.

Primary productivity and the biomass of primary producers are very unevenly distributed around the world. On a per-area basis, primary production varies by two orders of magnitude among habitats, from lows in Arctic tundra to highs in the coastal ocean and tropical rain forests. Today's biosphere fixes about 9×10^{15} moles of carbon per year.[5] The global biomass of terrestrial primary producers is estimated to be about 450 gigatons, whereas marine primary producers have a total mass of about 1 gigaton.[6] Although half of primary production occurs in the oceans, implying vastly higher rates of turnover there than on land, this fact masks crucial differences in the resource base on which marine and terrestrial life depends. The major realms of life thus differ in the types of organisms that fix organic matter and in the ways that these organisms acquire essential nutrients.[7]

In the oceans there is an important distinction between production that takes place in the pelagic zone, the open sea where single-celled phytoplankters floating in water are the primary producers, and the coastal belt where both phytoplankton and bottom-dwelling (benthic) organisms—seaweeds, seagrasses, and photosynthetic animals like corals—are at the base of the food chain. Productivity by phytoplankton is low in parts of the open ocean in so-called gyres where surface waters tend to descend, as well as around clear-water coral reefs. On the reefs themselves, however, benthic productivity can be quite high, because long-lived corals, sponges, and algae capture and accumulate nutrients. Further inshore from tropical coral reefs are highly productive mangrove forests and seagrass meadows, which contribute nutrients to the reefs themselves. Low planktonic productivity and high benthic productivity characterize coral islands of the Pacific and Indian Oceans. Reef animals show a conspicuous contrast in size between relatively small suspension-feeders, which feed on phytoplankton, and large herbivores that consume benthic primary producers. In the turbid waters along continental coastlines and in some atoll lagoons, both planktonic and benthic productivity can be very high, mirrored by the correspondingly large sizes and high abundances of both suspension-feeders and herbivores.[8] Shallow-water ecosystems on the Pacific coasts of Central and northern South America, the Caribbean coast of Colombia and Venezuela, and many temperate-zone coastlines exemplify this richness of the plankton and benthos well. These are the coastlines where suspension-feeding barnacles, mussels, oysters, and burrowing clams are large and abundant.

Intimate mutualistic associations between unrelated organisms have been instrumental in raising rates of primary production. Single-celled dinoflagellates, cyanobacteria (blue-green algae), and green algae have entered into close partnerships with suspension-feeding marine and freshwater animals including sponges, corals, giant clams, and sea squirts (ascidians), as well as marine benthic foraminifers and radiolarians, making effectively oxygen-producing plants.[9] Seagrasses, which are among the most productive plants in the coastal ocean, might have been unable to colonize sandy and muddy sediments without the aid of lucinid bivalves. These clams make sediments suitable for rooted seagrasses thanks to the presence of symbiotic bacteria in their gills.[10] Most land plants have fungi (mycrorrhizae) associated with their roots. The fungi aid the plant in creating soil and in making minerals from rock available to their hosts.[11] The highly diverse pea family (Fabaceae) comprises plants that have evolved intimate associations with root-dwelling nitrogen-fixing bacteria (rhizobia). By making inert nitrogen available to these plants in usable form with energetically costly enzymes (nitrogenases), the rhizobia have substantially increased the primary productivity of forests and croplands.[12]

There is a curious example in which a single-celled animal-like protistan, the ciliate *Pseudovorticella* sp., is symbiotic with a planktonic diatom. Like other algae, the diatom of the genus *Coscinodiscus* acquires nutrients passively at its cell surface. The ciliate increases the flow of nutrients to the diatom, and therefore its photosynthetic rate, by a factor of four to ten because its cilia create currents that bring more water to the diatom's surface. This association is found in waters where nutrients are relatively scarce and where the diatom by itself would not do well.[13]

Primary productivity and access to the means of production are important to all organisms because they set limits on how much power individuals, populations, and even ecosystems can wield. To put it perhaps more positively, increases in primary productivity enable some organisms to acquire more power. Positive feedbacks from consumers to primary producers ensure that the rate at which carbon is fixed is as much an enabling power as it is a constraint.

Predictability

To be most effective, the supply of resources as experienced by organisms must be predictable as well as abundant. An erratic supply relative to the life span of an organism renders the resource unreliable and therefore difficult to exploit.

An example from hermit crabs illustrates this challenge. Most hermit crabs live in and carry the shells of snails and other molluscs that have died; the shell serves as a mobile home in which the female carries eggs and in which the soft abdomen is well protected. The shell's opening can be closed by the crustacean's modified left or right claw (chela), which functions as a door when the crab withdraws into the shell. Hermit crabs compete intensely for shells, especially when they need to move into larger shells as they grow. A supply of new shells is necessary because worn-out shells, which the hermit crabs are unable to repair, become unusable. For many hermit crabs, a reliable supply of new shells is provided by predators of snails, at least those predators that leave the shells intact. Some hermit crabs, in fact, wait at predation sites and quickly take possession of a shell when one becomes available after the predator abandons it. Exploitation of the shell resource is feasible as long as predators release a steady supply of shells, but in many situations where predators are scarce, the availability of shells is unpredictable, akin to a boom-and-bust pattern. In rivers that flood from time to time, for example, shell-bearing snails die in episodes of mass mortality, flooding the market and making it difficult for secondary shell-dwellers like hermit crabs to exploit the sudden bounty. Much of the resource therefore goes to waste, at least from the hermit crabs' perspective. The erratic pattern of shell supply might account for the absence of hermit crabs in freshwater habitats. Some African cichlid fishes in lakes brood their eggs in abandoned shells, a specialization that is perhaps made possible by the abundant predators of snails in these lakes.[14]

Rather than depend on other species to make a resource more predictable, many plants and animals have found ingenious ways to store nutrients, either within their own bodies or at sites outside the body. As noted in chapter 1, plants store accumulated nutrients in underground bulbs, corms, and tubers as well as in above-ground stems. Some cold-water kelps (brown algae of the order Laminariales) accumulate nitrogen in basal parts of the plant during winter for use in the summer when nutrients are less readily available.[15] For land plants growing in warm, dry climates, photosynthesis is curtailed during daylight hours in part because of high rates of water loss. They take up carbon dioxide at night, when humidity is higher, temperatures are lower, and transpiration is less problematic. The carbon dioxide stored in the leaves at night is then used in the daytime for photosynthesis without undue metabolic cost.[16]

Internal storage as a long-term means of making an organism's nutritional supply more predictable is most effective in large organisms. Microbes contain vacuoles in which potential reserves are stored, and seeds and eggs contain enough nutrients for the embryo to develop; but even here, large size makes for greater storage capacity.

The largest stores, however, are kept in safe places outside the body. Jays, squirrels, and smaller rodents cache nuts, and acorn woodpeckers hoard acorns in self-made tree holes. Nut-bearing trees like oaks may even have come to depend on caching animals to disperse their seeds and place them in sites favorable for germination.[17] External storage is even known in some marine crustaceans such as callianassid ghost shrimps, which collect dead plant material in below-ground chambers.[18]

For some social insects like leaf-cutter ants, as well as for humans, there is a fine line between storing large quantities of food outside the body and cultivating edible organisms in well-tended farms. Cultivation of plants and fungi has evolved in ants, termites, bark beetles, bees, and humans; and some marine damselfishes (Pomacentridae) and limpets (Patellogastropoda) maintain territories in which they grow algal foods and keep out unwanted algal competitors.[19] The human species is unique in having extended storage and agriculture to include fermenting, canning, drying, and freezing food for consumption when fresh food is less available.

These examples illustrate an important aspect of supply and power. Simply having a large supply of nutrition in an organism's environment and having that supply come to the organism is not enough. The organism must do work not only to gain access to the resource, but also potentially to enhance that resource and make it more reliable. Access requires investment in time, in technology, and in power; it requires coping with competitors, including predators, which have the capacity to deny access.

To illustrate the importance of expending power to gain power, consider the case of suspension-feeding animals. Some of these, such as crinoids, passively wait for food-laden currents to come to them. With outstretched arms, crinoids catch food particles and then transport them by means of cilia to the digestive system. Passive food collection like this works well if suspended food is plentiful and if the food-gathering structures are in the direct path of moving water. If a passive suspension-feeder inhabits a crevice or lives under a stone or in some other situation where the water is stationary, food will quickly be depleted even if the surrounding water still contains a high load of suspended particles. The solution is for the animal to generate its own currents, drawing water to itself by means of beating cilia or moving limbs.

Another solution is to move the entire body. In both cases, these so-called active suspension-feeders continually sample fresh sources of suspended food. These power-assisted means of feeding provide greater access to suspended food and enable animals to live in sheltered sites where they are protected from many kinds of predators.[20]

Rorqual whales such as the blue whale (*Balaeonoptera musculus*)—the largest animal living today—take this power-intensive filter-feeding to an extreme. Like other baleen whales, these giants feed on krill, planktonic euphausiid crustaceans that reach staggeringly high abundances in the productive waters of the Southern Ocean and elsewhere. During dives, blue whales gulp a vast amount of seawater, filtering out krill and other plankton through the baleen. According to Jerry Goldbogen and his colleagues, a 20-meter-long rorqual engulfs as much as 82 cubic meters of water, amounting up to 160% of the whale's body volume, by opening its mouth and moving forward. Despite the highly streamlined body of the whale, this lunge-feeding greatly increases drag on the body and is therefore energetically very costly. During a single dive, which lasts 15–17 minutes, each lunge nets about 12 kilograms of krill, and the whale can execute several of these lunges. Calculations reveal that daily energy demands would require 83 lunges, yielding an average of 11 kilograms of krill per lunge. Together with spring-like tendons associated with the jaw points, this extremely expensive style of filter-feeding enables rorquals to exceed all other baleen whales in body size. Bowhead whales and some other whales that filter while continuously swimming can also attain large sizes, but their method of filter-feeding, like that in plankton-feeding sharks, is still not as effective as the lunge-feeding of rorquals.[21]

This example highlights another important facet of the relationship between animal power and environmental production. A high abundance and productivity of oceanic plankton enables plankton-feeding whales and sharks to achieve gigantic proportions, but this food-rich environment by itself is insufficient to make giants out of medium-sized animals. Two additional factors are likely to contribute to gigantism. First there are agency and selection. The largest living planktivores are either fully warm-blooded (as are whales) or regionally warm-blooded (as are sharks). This means that, as a result of competition, they can wield extraordinary power to exploit food over a very large area and volume of ocean. Slower animals with smaller ranges could live in equally productive waters but never achieve the sizes of the warm-blooded planktivores, because they cannot meet the power demands necessary for feeding on such a large scale.[22] Second, these active animals not only redistribute nutrients by increasing the rate of turnover of animal biomass, but they effectively recycle nutrients. Instead of drifting down to the ocean floor, the feces of large planktivores stay in the top two hundred meters of the ocean, contributing to productivity there and thus stimulating the conditions that enable very large body size to evolve. In a sense, the animals are fertilizing their own food.[23]

Animals have at least two other means of regulating and potentially increasing a predictable food supply. Both are counterintuitive in that they involve the idea that consumption of a living resource makes that resource more productive.

Consider the case of plants. Without consumers, plants senesce and die, leaving their remains to be digested by soil fungi and bacteria, which eventually transform the once-living plant into nutrients that the next generation of primary producers can exploit. This recycling process is sped up with a shortcut when animals ingest living plant tissues and digest them, often with the aid of microbes. Moreover, the feces released by the herbivores directly fertilize plants. High rates of consumption by herbivores, especially warm-blooded ones, stimulate faster growth and higher productivity in their food plants.[24]

The second mechanism, orchestrated through adaptation by the plants themselves, encourages herbivores to consume older, less productive parts of the plant while the more actively photosynthesizing parts, which are often well hidden, are left intact. Grazing mammals, for example, "mow" the grass, taking off the tips but leaving the actively growing parts near the base unscathed. Cold-water kelps appear to employ a similar method to counter intense consumption by Steller's sea cows (*Hydrodamalis gigas*), large mammals that are now extinct. In both cases, the plants have become extremely productive despite (and likely because of) intense herbivory by large, warm-blooded consumers.[25]

The mere movements of many marine animals also have important stimulatory effects on production. By irrigating marine sediments with oxygen, for example, burrowing animals potentially release phosphorus to environments where anaerobic life would otherwise have prevailed. Besides facilitating colonization of sediments by aerobic animals, burrowers pave the way for grasses in salt marshes, as very nicely demonstrated for burrowing fiddler crabs that stimulate the salt-marsh grass *Spartina* in eastern North America.[26]

Relationships like these, in which animal agency contributes to higher plant productivity, are probably very common. A particularly interesting example comes from Africa, where hyenas have the habit of cracking and ingesting the bones of their prey. When they defecate at specific sites, the bone fragments in the feces release phosphorus, a limiting nutrient for plants, and therefore increase the productivity of the ecosystems in which the hyenas live.[27] Whether predators stimulate the production of their animal prey has not yet been investigated. An argument against such an effect is that, whereas most herbivores eat only parts of the plant and therefore do not kill their victim outright, successful predation of animals often does result in death. Such finality would make it impossible for animals to become more productive by sacrificing old tissues. Nevertheless, a stimulatory effect on animal production might be found in animals subject to partial predation, such as sponges and colonial animals, or in those in which tissues are easily regenerated, as are the siphons of burrowing clams and parts of worms.

Heat and Cold

Though essential for life, nutritional resources for which organisms compete are only part of the environment in which competition takes place. Many factors that are not resources nevertheless profoundly affect the rates at which resources can be extracted and the power that competitors can wield. Together with resources, these additional enabling or constraining factors influence power. They include temperature, gravity, and the density and viscosity of air and water, among others. Each of these everyday enabling or constraining conditions affects what organisms can do. Unlike selective agencies, enabling factors affect every living thing in a given environment, even if organisms differ in the degree to which the factors facilitate or hinder activity.

Consider temperature, for example. Temperature, the mean kinetic energy of a molecule in a substance, affects nearly all chemical reactions, including those involved in metabolism and activity. The rule of thumb, to which there are many exceptions, is that the rate of a reaction approximately doubles for every 10 degrees Celsius rise in temperature. In the case of most plants and animals, temperatures above 30–35 degrees Celsius tend to be detrimental as proteins begin to break down; but for many microbes, life can proceed at much higher temperatures. The implication of the doubling rule is that, all else being equal, activity becomes much cheaper as temperatures rise. The activation energy required to start a reaction is lower and the time required for a given activity is shorter, implying an increase in power as temperatures rise and a decrease as they fall.[28]

A few examples will illustrate the importance of temperature in granting power and access. The predatory tiger beetle *Cicindela hybrida* in Denmark, which runs after its insect prey, can muster a speed of 0.023 meters per second at 20 degrees Celsius and as much as 0.042 meters per second at 30 degrees.[29] Handling time of prey was found to be independent of temperature.

Anyone who has held a hand in ice-cold water for long periods, as I did on an expedition to the Aleutian Islands of Alaska in 1987, will know that gripping force is severely compromised, because muscles contract more slowly and produce less power as the temperature declines. Some animals might be able to acclimate to the cold, but adaptive compensation is never enough to counteract the effects of low temperatures. This is why I like warm weather.

The pervasive effects of temperature are evident also at the ecosystem level. The consumption of plant leaves, for example, is far more intense in forests than in the nearby tundra of European Russia. Insects damaged three times as many leaves and consumed eight times as much leaf area in forest as compared with tundra woody plants. Insects removed about 4.34% of woody-plant leaves in the forest but only 0.56% in tundra plants. Moreover, leaf-mining—an important style of herbivory practiced by many groups of insects—was absent in the tundra.[30]

One problem that plagues all studies of herbivory and predation in modern ecosystems is the human-caused decimation of most large, metabolically active animals. Observed levels of consumption are likely to be unrepresentative of herbivory and predation during modern prehuman times. Green turtles (*Chelonia mydas*), for example, are still major herbivores in northwest Australian seagrass meadows, but elsewhere they have been severely depleted, prompting some scientists to argue that seagrass beds experience little grazing. Large marine predators such as sharks, groupers, toothed whales, and billfishes no longer exert the ecological influence they once did before intensive human exploitation.[31]

Such qualms notwithstanding, some credible studies show the expected pattern of more intense predation under warmer conditions. In one such study, a series of identical panels was set below the tide line at each of thirty-six sites spanning 115 degrees of latitude on both coasts of the Americas. Some panels were protected from fish while others were exposed to these predators. The results, reported by an international team of sixty-five scientists, showed that fish predation is much more intense in tropical waters than at higher, cooler latitudes in both hemispheres and in both the Atlantic and Pacific Oceans. The robust relationship between predation and temperature is the consequence of the larger size and greater mobility of predators and of the longer duration over which predation takes place.[32]

Rather than be at the mercy of environmental temperature, many animals and even some plants have evolved stable and often high internal temperatures. Familiar examples of such so-called endotherms are mammals, birds, and some insects. Some marine sharks, billfishes, tunas, and the deep-water opah (the fish *Lampris guttatus*) have achieved regional endothermy, in which some internal organs such as eyes and the brain are heated, and heat is retained by having blood recirculate deep inside the body. Compared with similar-sized ectotherms, whose body temperatures are not very different from those of the surrounding air or water, endotherms consume five to ten times more food.[33] Some large animals including the leatherback turtle (*Dermochelys coriacea*) achieve high and relatively constant internal temperatures by virtue of their very large size. Their modest internal heat production, coupled with a large volume relative to surface area through which little heat is lost, enables these large animals to generate enough heat to function as if they were true endotherms. This so-called inertial homeothermy is widely thought to have characterized large Mesozoic dinosaurs as well.[34]

As a means to attract and even trap pollinating insects, some flowering plants such as skunk cabbage (*Arisaema foetida*) and related members of the Araceae, as well as the large Amazonian water lily of the genus *Victoria*, raise the temperature of their flowers so that their attractive scents are copiously released. Such partial and temporary endothermy can be effective at night and even during cool late-winter months.[35]

The cost of doing life's business also crucially depends on temperature. The basal (or resting) rate of aerobic metabolism is the power that organisms must expend simply to maintain themselves. As temperatures rise, this rate increases exponentially while the solubility of oxygen in water declines, imposing a constraint on both activity and maximum body size. Many organisms ease this constraint by actively displacing oxygen-depleted with oxygen-rich water by moving part or all of the body.[36] An even more effective solution, adopted by many bony fishes as well as shore crustaceans and all land animals, is the ability to breathe air. Oxygen diffuses three hundred thousand times faster in air than in water and thus enables air-breathers to achieve much higher resting and active metabolic rates than animals that rely on oxygen dissolved in water.

Finally, a counterintuitive relation between mineral solubility and temperature affects the formation of protective mineralized skeletons. Ever since the latest Ediacaran periods, about 555 million years ago, organisms in dozens of independent lineages have evolved a skeleton as part of the body. This skeleton, made of minerals such as calcium carbonate, calcium phosphate, silica and other less common ones, provides protection against enemies as well as support for muscles involved in mobility. The solubility of calcium, silicon, and other ions decreases in water as temperatures climb, implying that the power required to build a skeleton, especially a thick, complex one, is much lower in warm waters than in the cold. It is possible in principle for a snail or clam in Antarctica to build a shell as robust as one on the reefs of New Guinea, but the cost would be prohibitively high, interfering with other essential functions. In practice this is never realized, because cold-water predators are much less powerful than their warm-water counterparts, so that selection for well-armored shells is weak or even absent.[37]

Air and Water

One truism that at first glance might seem unworthy of attention is the observation that plants and animals living on land differ strikingly from those populating the world's seas, lakes, rivers, and springs. This contrast is not just the direct consequence of differences in the physical and chemical properties of water and air. The differences have been enormously amplified by evolution, which through a combination of agency and natural selection has led to the exploitation of different capacities and modes of life that are compatible with the characteristics of the two media.

Compared with air, liquid water at 20 degrees Celsius is fifty-five times more viscous, has 25.6 times as much oxygen per liter, ten times higher diffusivity of oxygen and carbon dioxide, ten thousand times the rate of diffusion of these gases, less than one-fourth the specific heat, and 831 times the density. These ratios vary somewhat with temperature, but the pattern is consistent.[38] The consequences are numerous. Animals moving through water, and seaweeds and stationary animals exposed to water currents, experience far higher resistance (friction drag) than their equivalents in air. It takes far more power for an animal of given mass and velocity to move through water than through air. It should come as no surprise that the fastest swimmers—tunas and dolphins—have lower speeds (up to 15–20 meters per second) than the fastest runners (the cheetah *Acinonyx jubatus,* 25.9 meters per second) and fliers.[39]

One of the most consequential contrasts between air and water involves gravity. The average density of organisms is 1,073 grams per cubic meter, which is only 1.07 times greater than that of seawater. Organisms surrounded by water are therefore little affected by gravity. They can float or swim without sinking simply by buoying the body with metabolically produced gases, and even very large bottom-dwelling seaweeds such as the kelp *Macrocystis pyrifera,* which reaches a length of more than 45 meters, do not invest in costly support as plants on land do. The great abundance of organisms living unattached in water makes it possible for suspension-feeders to filter them with collecting devices. Comparable filter-feeding is uncommon on land, although some plants collect dust and other detritus from the air for eventual use as a source of nutrients, and spiders collect flying insects by snaring them in sticky webs.[40]

Marine organisms often disperse gametes and larvae in the plankton. Larval stages can spend weeks, months, or in some cases even years feeding in the plankton before settling to the seafloor, where they become adults. Because of gravity, this option is largely unavailable to the gametes and young stages of land animals. Spore-bearing and wind-pollinating plants, however, do rely on passive dispersal; and small insects and even some birds, such as the European swift *Apus apus,* can spend weeks to months in the air without landing.[41] On land, animals are responsible for the dispersal of most spores, pollen, and seeds. This animal-assisted dispersal is all but unknown in the sea.[42]

Agency and Environment

How much agency an organism has depends on external conditions as well as on the organism's structure and physiology. Greater agency is enabled (though not always realized) when productivity increases, temperature rises (at least up to about 30–35 degrees Celsius), and organisms live in air as compared with water. Atmospheric and ocean-water chemistry are also important enabling factors: Life can do a lot more things in the presence of oxygen than in its absence. Everything from sensation and communication to locomotion, passive defense, metabolic fate, body size, and the frequency of competitive and predation-related interactions is affected by these enabling factors. All these conditions, and the agency they permit or constrain, are subject to modification by and adaptive evolution of living things. How such changes can be inferred from the historical record, and what that record reveals about the evolution of life's agency, is the subject of the rest of the book.

3

Patterns of Power

HOW MIGHT THEY BE DETECTED?

In chapter 1 I suggested that the most powerful organisms and ecosystems exert an outsized influence on the distribution, adaptive characteristics, activity, and evolution of life. Given that power is, with few exceptions, difficult to measure directly in systems of the past, it is important to ask what can be known and which methods yield credible answers to questions about the history of power. If power at many levels of life's organization did increase over time, which specific predictions can we make—and test—to bolster or quash this claim? Equally important is the question, which lines of inquiry are uninformative, sterile, or even misleading. Put another way, what do we need to know, what can we know, and which kinds of evidence are most pertinent to uncovering patterns of life's power in space and time?

Any such inquiry must begin with the identification and characteristics of the most powerful living organisms. Who are they, and how fast, big, strong, responsive, perceptive, and productive are they? How do they interact with other living entities? What happens when these potent agents are added to or eliminated from an ecosystem? We can ask similar questions about the most powerful ecosystems. What are their characteristics? How do they interact with other systems, and how do they recycle nutrients? Once such questions have been answered, inferences can be made about organisms and conditions in the distant past, many of which will be profoundly alien to our modern sensibilities.

Crucial in this evolutionary context is determining under which conditions power-enhancing innovations arise and spread, which circumstances might cause them to disappear, and how life itself has affected the prevalence of these permissive conditions.

The power-related performance of organisms from the distant past must be inferred with knowledge of the properties of living species and the principles that link these properties to performance. With that information in hand, the next problem for the historian of power is learning how to recognize and document the distribution of power over time. Scientists usually resort to statistical analyses to distinguish a meaningful signal from a false trend that reflects mere random fluctuation, insufficient sample size, or some other hidden artifact. The question is this: How do we know whether an apparent trend in maximum power reflects biological and economic reality?

The crux of the argument in this chapter is that it is misleading to assess patterns in maximum power by considering the performance of all species, that is, by treating powerful species as part of the larger sample of species that includes much less powerful ones. Instead, I suggest that comparisons must concentrate on one end of the spectrum of power and on how the composition and performance of agents at that end of the distribution change over time.

Powerful Statistics

The most powerful entities are by definition at one end of a statistical distribution of performance, a distribution that also includes much less powerful agents. In fact, there are very few extremely powerful entities, but legions of less potent ones. To explore the performance and influence of these most powerful agents, therefore, it makes no sense to rely on measures of central tendency such as the mean, median, mode, variance, or standard deviation of power-related traits. If, for example, we tracked body mass of the most powerful mammals over time, we would avoid using the average body mass of all mammal species at any given time. Summary statistics like the mean, median, or mode incorporate heterogeneity and variation, and treat all data points as equals, even though it is the organisms at the extreme end of the distribution that wield the greatest influence.

A thought experiment will make my reservations about the reliance on central tendency clear. The human species represents one among millions. As such, its evolution would not have changed the mean or any other measure of central tendency for any power-related trait when the overall statistical distribution of that trait is considered, whether it be for animals as a whole, for all mammals, or even for all primates or all hominins. Yet this singular event represents an unprecedented increase in power, one that ultimately changed Earth's geology and biosphere. Statistical significance, in other words, can be a poor guide to biological significance.

As another example, consider the consequences of past mass extinctions. Most studies of these events rely on tabulations of lineages—species or more commonly genera—and their temporal ranges as revealed by their distribution in the rock record. Some lineages survive the crises while others succumb to the catastrophic disruptions. To assess whether the extinctions are selective with respect to some trait like size or locomotor performance, scientists assign a value to each lineage and then determine whether surviving lineages differ with respect to that trait from lineages that became extinct. The assumption underlying this kind of analysis is that all lineages are equal. This is a reasonable assumption if the aim is to assess how lineages belonging to different major categories behave during the mass extinction, but it can be more problematic when applied to performance. If, for example, larger-bodied lineages are more powerful than smaller-bodied ones, the loss of even one lineage of giants would have greater economic repercussions than the loss of a lineage of small animals. Yet the disappearance of that most powerful entity might not be detected statistically and would not be regarded as statistically significant. The loss of many small-bodied lineages would obviously also matter to the system in which they reside, for it might point to factors that make those lineages especially vulnerable; but treating all lineages as equals with respect to the consequences of the loss of powerful agents could yield misleading or even erroneous conclusions.

Concentration on the extreme end of the distribution of performance does not mean that the rest of the distribution should be ignored. It is informative to ask, for example, how much power differential exists between the high and the low end of the power distribution; and which proportion of entities exceed or fall below a given level of performance. It is also important to ask whether the distribution as a whole is displaced toward greater power, or to ask whether species formation occurs more rapidly in one part of the power spectrum than in another. These and other questions require the whole distribution of performance levels. In studies of inequality of wealth and income in human societies, for example, it is common and useful to compare the top quartile, decile, or centile with their bottom equivalents, or to establish how those top units behave geographically over time. This is the approach Thomas Piketty used in his wide-ranging studies of inequality.[1]

It has become routine in some quarters to ascribe changes at one end of a distribution to random processes or to the effects of sampling size or sampling intensity. This trend is again well illustrated by studies of body size. In most major branches of the tree of life, the ancestor is small, as in the origins of vascular land plants, seaweeds, molluscs, birds, and mammals, among many others. Subsequent evolution is more likely to extend the distribution of sizes toward larger organisms than toward smaller ones because there is, metaphorically speaking, more room to the right than to the left of the ancestral size. The observed increase in maximum size over time could thus be regarded as a statistical artifact, the result of a random walk among evolutionary lineages. In this interpretation, no causal mechanism need be invoked to explain the expanded size range of the group.[2]

Biological philosophers Daniel McShea and Robert Brandon expanded the idea that no explanation is necessary for a pattern of increasing diversity through time by proposing what they called the zero-force evolutionary law (ZFEL). Selection, they argued, is unnecessary as an explanation for diversification in lineages over time because, in the aggregate, members of the lineage act like molecules in a gas. Selection and other potential forces acting on lineages might exist, but if random movement suffices to describe a collective pattern, such forces are simply superfluous.[3]

From my perspective as an organism-oriented evolutionist, the problem with this line of thinking is that the random-walk interpretation describes a change in the whole distribution and does not consider the behavior of individual points in the distribution. These points, if they correspond to organisms or evolving lineages, respond to and cause effects that are unique to them. In the aggregate, these causes and effects could be described as random, but a focus on individual particles reveals a key distinction between inert, identical molecules in an idealized gas on the one hand and living, evolving agents on the other. Confusing aggregate behavior in a distribution of heterogeneous living things with the behavior of particular agents and their adaptive evolution leads to interpretations that obscure meaningful variation and difference. Agents are adaptive wholes with activity. Unlike interchangeable molecules, they have some input in determining their future. Those agents would not exist without adaptations, which are the consequence of selection and agency acting together. Agency is an inescapable and distinctive attribute of living things, and as such it violates the key assumptions for random models that agents are interchangeable and unchanging particles and that, if agents are alive, the forces acting on them cancel out in the aggregate.

My criticisms should not be taken as an assault on statistics. Rather, they are meant to draw attention to the limitations of statistical inference and the suppression of economically and biologically meaningful variation. An overreliance on arbitrary levels of statistical significance can lead to the acceptance of conclusions with little or no real-world importance and the failure to identify consequential phenomena that are detectable only at the extremes of a distribution. Conclusions drawn from any kind of analysis are only as good as the data and assumptions used. If analyses are faulty, the consistent inferences drawn from them are not necessarily the right ones.

Diversity and Power

Another concept that summarizes variation and that has often been linked to studies of adaptive evolution is diversity, usually defined as the number of species or lineages. It is widely assumed, but rarely demonstrated, that adaptive breakthroughs, including those that yield an increase in power, lead to subsequent diversification in the lineage in which the innovation arises. By its collective nature, however, diversity expresses variation and difference, not adaptation or power or any other trait of individual organisms. It encompasses species that vary in competitiveness, power, and performance. High-powered entities comprise a minority of lineages in any ecosystem just as powerful people and institutions are greatly outnumbered by those with less power in human societies. The total number of entities in a system, or for that matter in a major evolutionary branch of the tree of life, should therefore be a poor reflection of the power of that system or of its most powerful members. Even the proportion of highly potent species relative to less powerful ones above or below some threshold is an unreliable measure of power. On land, for example, the great majority of animal species are non-social insects, which are small to very small creatures whose individual and even collective power pales in comparison with that of the much smaller number of social insects and larger vertebrates with more power.

To make this point clear, consider a hypothetical case. If, in the aggregate, the number of species of non-social insects increases much faster than the number of social ones in an ecosystem, one might conclude from measures of diversity alone that the system overall has decreased in power. But such a conclusion rests on the assumption that species are interchangeable relative to power. Social insects may not have risen as quickly in their diversity, but they could have gained in power.

Diversity is nonetheless profoundly important when viewed from an economic perspective. Instead of considering species only as phylogenetic units that are genetically more or less distinct from each other, a species can be thought of as unique in its ecological role. In this interpretation it is defined by what individuals do, how and where they make a living, and how they interact with enemies and allies. There is often a degree of division of labor within a species, as exemplified by the different modes of life of larvae and adults or in castes of social insects, but individuals affect the lives of others and of the system in which they compete. This ecological view of species is economically equivalent to occupations in human society. Like species, occupations exemplify division of labor, without which the economy would collapse.[4]

Diversity as an expression of division of labor exists in all living systems from enzymes and organelles within a cell to tissues and organs in a living body and species in ecosystems. Division of labor—that is, specialization to particular tasks or functions—arises ultimately from competition among parts. Small differences in performance among parts can result in a selective process in which specialization is favored. This process rests on the reality that no part of the system can do everything equally well; consequently, tradeoffs in performance among different tasks result. Diversity in this sense therefore comes about through competition; it is not simply an expression of accumulating random errors that by themselves are irrelevant to how well functions are carried out.

Insofar as it reflects division of labor, diversity in this economic sense is an essential property of ecosystems and human societies. These larger systems could not exist without the roles that producers, consumers, and regulators fill. In the case of ecosystems, moreover, as each functional category is represented by more species, such as primary production and decomposition, the more robust and resilient the system is to disruption, because the loss of one or a few species will be compensated for to some extent by other similar species. Diversity, in other words, provides redundancy of function as well as unintended cooperation among members with complementary roles.

There is one other way in which the power of ecosystems reflects, and is reflected by, diversity and diversification. This involves sexual selection and its role in the formation of species. Most land animals and many plants engage in explicit mate choice, which is accomplished by a wide variety of costly signals and displays. Small changes in mating-related traits can cause lineages to become isolated and as a result to diverge as separate species. Mate competition is therefore a powerful engine of species formation, one that enhances and is made possible by substantial power. The combination of intense mate selection and susceptibility of lineages to species formation has resulted in extraordinarily high diversity of land plants, arthropods, and vertebrates. There is, in other words, a potential evolutionary feedback between elaborate mate-related selection and ecosystem-level power.[5] Signals and displays are also common in marine animals, but they tend to work over shorter distances because of the properties of water, and many marine species do not practice individual-level mate choice. This may be one of several reasons for the lower diversity and less power observed in marine as compared with terrestrial ecosystems.

In short, the concept of diversity becomes economically meaningful when it is applied to function and the division of labor. It is an emergent property of systems, the consequence of many interactions and long-term processes ranging from day-to-day competition to the evolutionary formation and disappearance of species. By itself, however, diversity cannot be taken as a substitute for power or a measure of adaptive evolution.

Predictions About Patterns of Power

To be useful as a coherent theory, an evolutionary theory of power must account for known facts as well as imply predictions about phenomena that are not yet well recognized. The broadest prediction is that power increased over time during most time intervals and in all economic systems because of relentless selection and agency to maintain or increase power under the circumstances in which individuals live. For the most potent entities, the height of successive peaks in power should increase over time. Specific predictions targeting particular dimensions of power flow from this more general expectation. These dimensions include body size (length or mass), distance and velocity of movement, force employed in feeding and defense, rates of metabolism and growth, and for ecosystems, the rate of turnover of biomass. Importantly, these patterns should apply, though with different limits, to all economic systems of living things.

Another central prediction is that powerful entities force less powerful ones into the less productive parts of ecosystems or societies, where innovation and increases in power should also be expected. Initial refuges should therefore gradually but consistently become places of more intense competition as ecosystems and the biosphere as a whole expand.

This process should result in greater autonomy of the entire system from external disruptions. Moreover, if disruptions do compromise the most powerful entities, less powerful organisms would be in a position to replace them and to become the next powerful hegemony. Ecosystem and biosphere expansion powered by adaptive evolution should increase power even in the most productive parts owing to the potential for evolution to replace failing incumbents.

Increases in power must often depend on innovation or invention. Endothermy, the neuromuscular system, plant roots and leaves, and symbioses between organisms are well known examples in evolution; fire, agriculture, engines running on fossil fuels, and computers are obvious examples in the human realm. As an empirical matter, it is important to identify these breakthroughs, to assess whether they did indeed lead to an increase in power, and to specify the conditions that enable and select for them. Long-term increases in power likely require that breakthroughs accumulate and build on each other either in the same lineage or among different players in the whole system. The resulting economy and innovation-generating infrastructure should relax limitations and create opportunities for power-related innovations to proliferate.

At the scale of ecosystems and the biosphere as a whole, we should observe an overall increase in recycling of essential materials. If materials are lost to the system, the latter cannot be sustained for long. Effective recycling entails a profound level of cooperation among competing life-forms, a scale of facilitation that itself expands in space and over time and that regulates, however diffusely, the activities of even the most powerful entities. The scale and intimacy of cooperative arrangements in ecosystems should therefore increase over time, along with diffuse regulation. In other words, these positive feedbacks favoring the common good should increase over time. Natural selection, activity, and enabling factors are all involved in these feedbacks. They are interconnected and, if power did indeed increase over time, should be mutually reinforcing.

Finally, the history of life as a whole and of our own species is full of disastrous disruptions: mass extinctions in the case of the biosphere, wars and climatic calamities in the case of human societies. If power at all levels shows an upward trajectory, these disruptions should not have wiped out all the gains, but instead represent brief setbacks from which recovery and additional gains in power are rapid. Extinction of lineages is therefore not the same thing as the permanent extinction of power-enhancing innovations or of ecosystem feedbacks. Individuals, species, ecosystems, and social institutions come and go, but the fundamental traits of these entities survive, or are rapidly re-evolved, from a resilient base.

Escalation by Descent and Replacement

If these patterns are verified, questions arise about which evolutionary processes bring them about. For individual-level power-related traits, two evolutionary mechanisms are available: adaptation within lines of descent in a process of escalation in response to enemies, and replacement of one lineage by another. The second explanation cannot occur without the first, implying that within-lineage increases in power must be very widespread. This does not mean, however, that power increases over time in all lineages, or even that lineages should show consistent trends over time. For one thing, some species in a lineage may be unable to gain power because they do not prevail in competitive bouts and therefore retreat and specialize to refuges where competition is less intense. For another, organisms at the high end of the power spectrum could be affected by an external disruption or by agents that compromise their position as top competitor, leaving the door open for other lineages to come in and eventually usurp their incumbent status. Both descent and replacement are thus important, but the former should be especially prominent over the short term, whereas usurpation or more gradual replacement should be evident on longer timescales. The important methodological point is that phylogeny—the pattern of descent and branching within lineages and within the branches (or clades) of the tree of life—should be used in combination with cases of replacement in the fossil record to test whether escalation and increases in power have taken place.

An example from the history of American land vertebrates will illustrate the role of both descent and replacement. During much of the

Cenozoic era—the last 65 million years—the position of apex predator was held by borhyaenid marsupials and flightless phorusrhacid birds in South America; both groups are now extinct. Size increased over time in both groups, as well as in many of the herbivores on which they fed. Their hegemony ended when more powerful carnivores such as dogs, cats, bears, and weasels arrived from North America beginning about 9 million years ago during the Late Miocene and accelerating after 3.5 million years ago as a continuous land bridge with North America was established in the Central American isthmus during the Pliocene. The North American predators proved to be superior competitors, due in large part to the long history of intense competition on that much larger continent, which was also connected to the even larger Eurasian landmass for much of the Cenozoic. Escalation and within-lineage evolution in response to competition occurred on both American continents, but the invasion of South America by North American mammals represents a clear case of the usurpation of incumbents.[6]

The prevalence of replacement over inherited power applies equally in the human realm. Historians have long observed that great empires last only a few centuries at most; usurpers from outside take over as internal divisions and sometimes natural calamities take their toll on the incumbent seat of power. Even in those states in which the leader is a king or queen whose position is inherited, replacement is frequent. In the nearly one-thousand-year history of the British monarchy since 1066 CE, replacement as compared with inheritance of the royal mantle (counting siblings as cases of inheritance) occurred at least four times. Since the year 221 BCE, when China was first united, there have been eight dynasties including the Communist state since 1949, the only dynasty not based on inheritance.

It is important to emphasize that replacement of an exceptionally powerful entity by another need not be the direct consequence of competition between these agents. As Michael Benton and others have pointed out, there can be a gap of millions of years between the demise of one powerful species and the appearance of a comparably powerful one. Large herbivorous and predatory dinosaurs, for example, became extinct 66 million years ago, at the end of the Cretaceous period and Mesozoic era, and were not replaced by relatively powerful mammals until 9 million years later, during the Late Paleocene epoch. Similarly, lineages of very large clams whose tissues contain either photosynthesizing dinoflagellates or sulfide-oxidizing bacteria repeatedly became extinct, but they were not replaced by similar-sized or larger members of other lineages for 10–20 million years. Gaps of similar magnitude characterize the history of gigantic mobile pelagic plankton-feeders. Even in human history, there are often gaps of centuries between the demise of a powerful empire or dynasty and the emergence of new ones following a period of instability and chaos. The Holy Roman Empire under Charlemagne emerged 357 years after the fall of Rome in 411 CE. Chinese history likewise chronicles long periods of instability before the establishment of the Han and Ming dynasties.[7]

If powerful entities disappear for reasons other than competition with powerful rivals, their replacement much later with a new hegemony of powerful agents must imply that ancestors of the next seat of power must already have possessed traits that predisposed descendants to evolve into the power vacuum. What these properties are and why the time lag can be so long are topics for later chapters. Here my point is that replacement, with or without a time gap between successive, maximally powerful entities, is as important a phenomenon as is direct phylogenetic descent.

Powerful entities come and go, but the institutions or ecosystems over which they preside often have great staying power. Christianity has existed for more than two thousand years despite its many upheavals, schisms, and heresies. Its leaders have only occasionally inherited their position from their predecessors. Nation-states such as China, Great Britain, France, Russia, and Turkey have existed much longer than their times of peak power and extent of empire even though their borders have undergone considerable change. The same can be said of the broad categories of how organisms make a living: herbivory, predation, suspension-feeding, and the like. Reefs, forests, the level-bottom seafloor, and the pelagic realm persisted even when the composition of their biota underwent dramatic revolutions.[8]

The implication of short-term escalation and long-term replacement is that the most robust historical trends in power should emerge over long timescales. The cumulative nature of power-enhancing innovations, together with infrequent major biosphere-scale disruptions, geographical changes in boundaries between regional biotas, and of course the eternal pursuit of power should yield an overall trajectory toward higher maximum power and an expansion of the domain of life.

A Philosophy of Evidence

All this emphasis on methodology and predictions may strike some readers as overly academic and fussy. I would counter by saying that the conclusions we draw from evidence are only as good as the methods we use and the quality and reliability of the data we gather. For the study of power and its history, it is often not possible to measure or estimate performance directly, particularly when it comes to entities of the past. For this reason we must resort to proxies, measures that correlate with or indicate power. These proxies, and the analytical techniques we choose to document and interpret them, must be carefully evaluated. Science is often drawn to techniques and to statistical analyses that can be repeated or replicated, offering the comfort of robustness; but if the techniques are misleading or based on faulty assumptions, they may cause us to draw incorrect or irrelevant conclusions no matter how repeatable they are.

There is also a worrisome tendency to rely on analytical techniques that are so complex that their assumptions remain hidden. Often such complex methods, enabled by powerful computers, manipulate data in ways that obscure the quality and reliability of the data. For this reason, I tend to favor simple statistics and as little manipulation of the original data as possible. Clearly, there is a place for what are called megadata, the vast quantity of data that only a machine can organize and analyze. Genomes, languages, isotopic values at thousands of sites, and surveys of very large populations and economies require sophisticated algorithms for analysis, but many other kinds of evidence do not. For the patterns I discuss in this book, the original data often come from technologically elaborate estimates, modeling, and measurements, but their analysis does not demand heroics.

None of this should be construed as skepticism about methods properly applied. I care deeply about methods and numbers and reliability, and I strive for precision to the extent that the quality of the data allows. It is indeed out of respect for methods and for inferences drawn from them that I approach my studies and those of others with a keen awareness of the limitations and assumptions inherent in the work. In the remainder of this book, therefore, I proceed with caution and reasoned skepticism but also with the expectation that the record of the past can reveal fundamental truths about the nature and history of power.

4

The Evolution of Size

It is hard to imagine how different life on the early Earth was from the organisms and ecosystems so familiar to us in the present day. There were no trees, no flowers, no singing birds or chirping crickets, no whales in the sea or elephants on land, and—worst for me, perhaps—no shells on the beaches. Until about 650 million years ago, almost all living things were microbes, single-celled or very simple multicellular organisms that lived in the pelagic realm or formed mats on the seafloor and on land. Except for toxins and venoms, most of life's chemistry had evolved by this time, but the structural and behavioral dimensions of organisms were still in their infancy. It was a silent world devoid of the fragrance of living plant life. It was a world that knew no fear, where organisms moved little or not at all and where eyes and brains lay far in the future.

The story of how agency and natural selection transformed the microbial world into the diverse, sensually rich, and economically and technologically sophisticated biosphere of today is not simply a litany of names, numbers, and turning points. Rather, it is above all a story of functions and forms, of how feedbacks between organisms and Earth's environment lifted constraints on metabolism and created opportunities to evolve previously unimaginable living states as life's influence grew and adaptive breakthroughs proliferated and accumulated.

To tell this story, I begin with the power-related properties of individual organisms. I shall show that, despite occasional and sometimes long-lasting setbacks, power increased among the most powerful entities along every dimension and in every category of performance, from body size to movement, metabolism and defense. The natural history of power in the past as informed by the lives of modern organisms is the principal source of evidence on which this account rests.

Much of my emphasis in this chapter is on maximum power because, as I discussed in chapter 1, organisms with the highest performance levels have the greatest influence on their surroundings. I shall review power in less influential life-forms as well, for these organisms too have substantially affected their ecosystems and often enabled top competitors to achieve their power. By concentrating on the extremes, however, I draw attention to what is possible at different times given the prevailing mix of enabling factors and top-down controls.

The most obvious dimension of individual power with which to begin this inquiry is body size, the most passive manifestation of performance. Not only does it capture many aspects of individual power, but it is also relatively easy to measure or infer. Some degree of estimation is almost always necessary, especially in fossils, because body mass—the ideal measurement—must be inferred from more straightforward linear dimensions such as length of diameter. Nonetheless, the history of size offers a rich source of insights into the evolution of power.

Large size often confers a competitive advantage to its bearer. It enables individuals to apply more resources to interactions, to achieve a size refuge from predators that often take smaller prey, and to achieve a degree of internal homeostasis, or independence, from external conditions, as long as large size is associated with a small surface area compared with internal volume. Large bodies therefore passively establish an internal physiology that goes some way toward buffering them from external fluctuations.

As a dimension of power that is principally characterized by the accumulation of energy (itself a dimension of power), maximum size as a measure of maximum performance might be expected to be one of the earliest attributes that powerful organisms emphasized. More active means of achieving high performance, and requiring a more continuous supply of energy-rich resources, might be expected to reach maximum values later than the more passive dimension of size. The one exception to this prediction would be if size is expressed as the sum of body sizes of coordinated social groups. Individual organisms in such groups might be smaller, but the societies they comprise could attain all the benefits of individual large size by spreading performance among many individuals that, by virtue of effective communication (itself energetically costly), could spread power over a much greater space and a broader range of resources than could gigantic individuals acting alone.

From Microbes to Mammoths

Organisms today span some eighteen orders of magnitude in body volume. On the longest timescale, maximum body size jumped in two great pulses, separated in time by about 1.3 billion years.[1] The first occurred perhaps as early as 1.9 billion years ago, when an intimate association was established between two single-celled prokaryotes, an anaerobic host archaeon, and an oxygen-using guest proteobacterium (the future mitochondrion) to form the eukaryotic cell. Maximum size increased by about three orders of magnitude from prokaryotes to eukaryotes. The largest bacteria reach 750 micrometers in diameter, whereas the largest known single-celled eukaryote with a single nucleus (the benthic foraminifer *Nummulites* from the Middle Eocene of the Paris Basin of France) attains a diameter of 120 millimeters. Single cells with multiple nuclei—so-called coenocytes—grow much larger, as in some seaweeds and many fungi. The average mass of a living bacterium is about 2.6×10^{-12} grams, whereas the average eukaryotic protistan cell has a mass of 4.0×10^{-9} grams.[2]

Even more dramatic was the expansion in metabolism and in the number of genes expressing the synthesis and regulation of proteins. The eukaryotic cell represents a 200,000-fold increase in genetic expression compared with the prokaryotic condition. The per-capita metabolic rate rose by a factor of almost 5,000 during this transition.

The second phase began at about 650 million years ago during the Cryogenian period of the Neoproterozoic era, a time when all organisms were still soft-bodied. Eukaryotic algae became common, and animals achieved multicellular construction after passing through the single-celled zygote (fertilized egg) stage. Bodies composed of many interconnected cells evolved from single-celled ancestors perhaps as many as one hundred times independently, but animal multicellularity, characterized by functionally distinct tissues and thus by division of labor among parts, is thought to have arisen just once, beginning in a sedentary sponge-like marine form. By the Late Ediacaran, sponges capable of pumping water through choanocyte-lined chambers to extract nutrients already existed, as indicated by a study of the ratio of the areas of outflowing to inflowing currents through pores in the body.[3]

The time of origin of sponges and their pre-sponge ancestors remains contentious, with one report indicating a date as early as 890 million years ago. A tentative consensus, based on nonskeletal remains, is that the earliest centimeter-scale animals arose near the end of the Cryogenian, around 635 million years ago, but recent reports suggest that chemical signatures thought to indicate the presence of sponges during that time actually represent algae.[4]

By about 571 million years ago, near the beginning of the Ediacaran period, and perhaps only 4 million years after their first appearance, rangeomorphs—multicellular colonial animals thought to be stem-anthozoan cnidarians but lacking muscles, tentacles, and a mouth—had achieved a height above the seafloor of 2 meters. Within an interval of not more than 64 million years, therefore, maximum linear dimension of early animals increased by a factor of at least 200.[5]

Regardless of its precise timing, the advent of animal multicellularity opened the door to a vast increase in sizes and shapes. Much of this expansion was concentrated over the interval of about 150 million years, from the beginning of the Cambrian period of the Phanerozoic eon and Paleozoic era (542 million years ago) to the Middle Devonian Period (390 million years ago). During this time, maximum body volume increased one-thousand-fold, with another one-hundred-fold increase from the Middle Devonian to the present day.[6]

The latter increase, spanning almost 400 million years, sounds insignificant when it is expressed as orders of magnitude, especially when that one-hundred-fold increase is set against increases in earlier episodes. Orders of magnitude, however, can be deceiving. In terms of absolute volume, mass, or power, an increase by a factor of one hundred starting with a volume of something like 10^6 grams is huge compared with a one-hundred-fold increase beginning with an organism the size of a bacterium.

Algae and plants underwent a similar dramatic increase in size. Although some simple filamentous red algae of the class Bangiophyta are known as far back as 1.2 billion years ago, multicellular algae with distinct organs likely did not appear until the Cambrian or shortly before, and seaweed size remained on the scale of centimeters for 600–700 million years. Very large algae such as kelps in the brown-algal order Laminariales, some of which reach a length of 45 meters (as in the northeastern- and southern-hemisphere *Macrocystis pyrifera*), are thought to have relatively recent origins, perhaps arising as late as 35–40 million years ago during the Late Eocene. The largest kelps are likely not more than 15 million years old. In the tropics, green algae of the genus *Avrainvillea* (order Bryopsldales) form enormous masses more than 20 meters in diameter and two meters thick on mangrove islands in Belize. Their time of origin remains unknown, but related green algae extend back to the Cambrian.[7]

Like seaweeds, land plants had a slow start in terms of stature. They did not achieve organ-grade multicellularity until the Late Silurian (about 420 million years ago) with the definitive origin of vascular tissues. At this time, herbaceous plants preserved in the Bloomsburg Formation of Pennsylvania grew to a height of about 30 centimeters. By the Middle Givetian stage of the Middle Devonian period, about 385 million years ago, there were trees eight meters tall. Twenty million years later, during the Fammenian (the last stage of the Devonian), some trees reached a height of 40 meters.[8] Two modern species, the California redwood (*Sequoia sempervirens*) and the Australian gum tree (*Eucalyptus regnans*), exceed a height of 114 meters. Trees of this stature may already have been widespread during the Late Cretaceous.[9]

It is remarkable that terrestrial fungi large enough to have formed fossil logs achieved enormous sizes very early in the development of ecosystems. In the Bloomsburg Formation, the probable fungus *Germanophyton ptygmophylloides* reached a height of 1.3 meters, but this was nothing compared with the Late Devonian *Prototaxites owenii*, which formed logs 8.82 meters long with six branches, each with a length of one meter. These organisms have been widely considered to belong to the ascomycete fungi. Whether they might have been lichens, partnerships between a fungal host and algal symbionts, is controversial. Some smaller fossils contemporaneous with Early Devonian *Prototaxites* in the Welsh Borderland of southern England almost certainly had symbionts, but no evidence for a partnership with algae has yet been detected in the larger specimens. In any case, fungi achieved gigantic sizes, perhaps exceeding those of contemporaneous vascular plants in the Middle Devonian.[10]

Size Evolution in the Age of Predation

Consumption of one organism by another is a key selective process that should favor larger body size in both the consumer and the victim when the consumer has the potential to kill its prey. For the prey, large size represents a morphological refuge, because predators tend to take one prey at a time and find smaller prey easier to catch and kill than larger ones. Competition among predators likewise gives larger individuals the advantage because greater size provides access to a wider range of prey sizes. In fact, these advantages of large size magnify when predation involves greater force. For predators that swallow their prey whole, the prey is typically smaller than the predator; but for those that can kill or dismember the victim outside the body, this limitation is relaxed, with the result that many large predators can take prey larger than they are. If predators take prey in bulk, selection for larger size should apply to the predators but not the prey. If, on the other hand, predators do not kill their prey outright—think parasites, for example—selection for large size should be concentrated in the prey but not the predator. The history of size of consumers and their victims is therefore one of evolving power relations between the parties, affected both by interactions such as competition among perpetrators and by enabling factors such as productivity, temperature, and habitat size.

Predation had already emerged in the prokaryotic world. Modern bacteria penetrate and consume other bacteria, strongly indicating the antiquity of these forms of consumption.[11] The oldest eukaryotes might also have been predators, or else predation evolved multiple times among single-celled eukaryotic protists before the origin of multicellular animals. Phagocytosis—the process in which one cell takes in another—may be the oldest form of eukaryotic predation, but the oldest fossil evidence of predation comes from small perforations in testate amoebae made by protists resembling vampyroellid amoebae 780–740 million years ago. Predators were also responsible for complete and incomplete perforations in the calcareous tubes of *Cloudina*, one of the earliest mineralized animals (perhaps a polychaete worm) living about 550 million years ago during the Late Ediacaran period.[12]

Early sponges and their ancestors, with very large surface areas relative to their volume, subsisted either wholly or in part on dissolved organic matter, as do some living sponges today; but they also may have collected individual bacterium-sized organisms.[13]

A much later form of consumption is suspension-feeding, in which particles are collected in bulk with filtering devices. Besides sponges, the earliest seafloor-dwelling animal to acquire this ability was the passively feeding *Tribrachidium*, which appeared as early as 555 million years ago during the Late Ediacaran period.[14] Plankton had evidently become abundant and productive enough at this time to sustain this mode of bulk predation.

By the beginning of the Cambrian, both predation and suspension-feeding were well established. With the advent of mineralized skeletons and their evolution in many lineages of animals, protists, and algae, larger predators evolved weapons to bite, crush, and perforate the skeletons of prey. Given that even the earliest skeletons show evidence of unsuccessful predation, and that healed injuries are known in Early Cambrian brachiopods and trilobites, it is highly likely that predators were instrumental in favoring the evolution of mineralized as well as organic-walled skeletons. I shall have more to say about this weaponization in chapter 6. In any case, whatever limits to size might have prevailed before the Cambrian, they were dramatically relaxed toward the end of the Ediacaran period and beyond.

Gigantism in the Marine Paleozoic

Maximum size attained successively higher peaks among animals of the Paleozoic era, extending from 542 million years ago to the end of the Permian period, 252 million years before present. Throughout the era, the largest animals were consistently marine. The largest animal of the Early Cambrian was a radiodont arthropod, *Anomalocaris canadensis,* a swimming suction-feeding predator of soft-bodied prey. This species may have attained a length of one to as much as two meters. Seafloor-dwelling trilobites (*Redlichia rex*), reaching a length of 25 centimeters, were capable of damaging skeletons of trilobites and perhaps other animals.[15]

Predatory vertebrates surpassed arthropods in body size in the Late Devonian, when the placoderm *Dunkleosteus terrelli,* which reached a length of six meters, was the apex predator of its day, some 365 million years ago. Some 50 million years later, that role was filled by a ctenacanthiform shark, possibly a regionally endothermic animal with a length estimated to be seven meters. Finally, in the Late Permian, the largest marine vertebrate of the Paleozoic was another fish, *Helicoprion,* reaching perhaps ten meters in length.[16]

Actively swimming plankton-feeders have always reached very large sizes, and those of the Paleozoic are no exception. Oddly, however, and in contrast to later ages, these pelagic animals did not exceed the largest predators of their time in size. The earliest of the large plankton-feeders was the radiodont *Tamisiocaris borealis* from the Early Cambrian of the Sirius Passet Formation of Greenland. According to Jakob Vinther and his colleagues, this animal reached a length of 1.2 meters or more. Later, in the Early Ordovician of Morocco, *Aegirocassis benmoulae,* a related hurdiid radiodont, exceeded a length of two meters. The next peak in size, reached during the Middle Ordovician, was occupied by the largest shell-bearing animal ever to have lived, an endocerid cephalopod with a straight shell nine meters long in which the animal's soft parts occupied the anterior third of the shell. Evidence that this cephalopod and related endocerids were ocean-going plankton-feeders is circumstantial, consisting mainly of the very high abundance of these animals in sediments far from shore and their lack of a jaw. In the latest Devonian, about 365 million years ago, the role of largest plankton-feeder was taken by *Titanlchthys*, a six-meter-long placoderm fish. It is curious that no large plankton-feeders have so far been recognized from the last 100 million years of the Paleozoic.[17]

Early giant animals on land were arthropods. They reached peak sizes during the Carboniferous, with arthropleurids—relatives of centipedes and millipedes—reaching more than two meters in length. The largest flying animal of the era was *Meganeuropsis permiana*, with a wingspread of about 71 centimeters or more. It was not until the Late Permian when predatory vertebrates reached a mass of two tons (2,000 kilograms). Herbivorous vertebrates likewise attained this mass during the Late Permian. All these sizes, as will be seen in the next section, are well below Mesozoic and Cenozoic maxima for their feeding group or locomotor type.[18]

Giant Individualists and Smaller Socialists

The general upward trend in maximum size continued through the end of the succeeding Mesozoic and Cenozoic eras in the sea, but only to the end of the Mesozoic on land. Did organisms reach a size barrier, or did something else emerge that made gigantism a less effective means of gaining and holding power? Given the fluidity of limits, I am not confident that we can answer the first part of that question, at least for animals; but the second part, concerning the diminished value of extremely large size, deserves an answer of yes.

Land-based animals are good examples for the case that gigantism has not always signified maximum power. Throughout the Mesozoic era, from the Triassic to the end of the Cretaceous 66 million years ago, herbivorous land animals have exceeded their Paleozoic counterparts in maximum size. By the Late Triassic, the dicynodont *Lisowicia bojani* from Poland, representing the major branch of reptiles that also includes the mammals, is estimated to have weighed nine tons and to have been 4.5 meters long and 2.6 meters tall. The early sauropod dinosaur *Antetronitrus* from the Late Triassic weighed at least 5.6 tons. The Early

Jurassic *Vulcanodon*, another sauropod, attained a weight of 9.8 tons, whereas the Late Jurassic sauropod *Brachiosaurus* is estimated to have weighed 56 tons. The largest land animal known thus far, the titanosaur sauropod *Patagosaurus mayorum*, reached the astonishing length of almost 40 meters and a conservatively estimated body mass of 69 tons. This remarkable giant, which I had the privilege to inspect in a beautiful museum in Trelew, Argentina, lived during the latest Albian epoch of the Cretaceous, about 102 million years before present, in Chubut Province, Argentina. Even during the Late Cretaceous, South American titanosaurs and the North American *Triceratops* still weighed more than any Paleozoic herbivore or Cenozoic mammal.[19]

It took 25 million years after the end-Cretaceous extinction of large herbivorous dinosaurs before a new regime of huge plant-eaters was established on land. These large mammals, weighing 1,000 kilograms or more, appeared in the Late Eocene, about 40 million years ago. Neither they nor their mammalian successors came anywhere close to the sizes of the Mesozoic giants. The largest living terrestrial herbivore—and largest living land animal—is the African elephant (*Loxodonta africana*), which reaches a shoulder height of almost four meters and an estimated weight of twelve tons. Many fossil elephants were considerably larger. The most gigantic is *Palaeoloxodonta namadicus* from the Pleistocene of Asia. Based on a femur that could have been 1.9 meters in total length, an individual is estimated to have stood 5.2 meters at the shoulder and weighed 22 tons.[20]

Unlike herbivorous dinosaurs, which were inertial homeotherms, elephants and gigantic rhinoceroses are full endotherms, whose dietary requirements would be 3–5 times greater than those of similar-sized titanosaurs. This rough calculation, together with the rough estimates of body sizes of dinosaurs and elephants, might imply that elephants were at least as powerful as much larger dinosaurs despite their smaller maximum size.

Theropod dinosaurs and some Late Cretaceous crocodiles were the undisputed apex predators of the Mesozoic. In the Late Triassic, *Dilophosaurus* had attained a weight of about 350 kilograms, but later forms grew much larger. The allosauroid *Ulughbegsaurus uzbekistanensis*, belonging to the family Carcharodontosauridae, lived during the Turonian epoch of the Late Cretaceous (90–92 million years ago) and is estimated from a maxilla to have been 7.5–8 meters long and to have weighed about one ton. The largest known predator on land of all time is the famous Tyrannosaurus rex from the latest Cretaceous (Maastrichtian epoch) of North America, weighing in at some 7.7 tons.[21]

The largest living crocodiles are the American *Crocodylus intermedius* and the riverine and partially marine Indo-Pacific *C. porosus,* both reaching a length of about 6.3 meters. These species are only about half the length of 11–12 meters reached by the Late Miocene South American *Purussaurus brasiliensis,* the Late Cretaceous North American *Deinosuchus riograndensis,* and the Late Early Cretaceous (Aptian-Albian stage) African *Sarcosuchus imperator,* all capable of killing large vertebrates. The African species is thought to have weighed eight tons.[22]

At an estimated mass of 470 kilograms, the Pleistocene South American saber-tooth cat *Smilodon populator* is almost 1.8 times heavier than the living carnivore, the Siberian tiger *Panthera tigris,* and 1.1 times heavier than the Pleistocene North American lion *P. atrox.* The Pleistocene South American bear *Arctotherium angustidens,* with a mass of about two tons, is even larger, but like many other bears (Ursidae) it may not have been a strict meat-eater.[23] It is worth noting in passing that the large South American mammals mentioned here have ancestors that migrated from North America across the Central American isthmus and were thus not part of the native South American faunas that flourished before the isthmus was established.

Flying animals, originating during the Middle Carboniferous about 324 million years ago, never attained the sizes of their ground-dwelling counterparts, but they too show an increase in maximum size to a peak during the Late Cretaceous, followed by a decline in the Cenozoic. The largest known insect—and largest Paleozoic flier—is the Permian *Meganeuropsis permiana,* with an estimated mass of 100–150 grams. Some 40% of this mass consisted of flight muscle. The largest living insect, the South American cerambycid beetle *Titanus giganteus,* is a poor flier and is much smaller, with a length of 167 millimeters and a mass of 38 grams.[24]

The largest flying animal known is the gigantic Late Cretaceous azhdarchid pterosaur *Quetzalcoatlus northropi*, whose wingspan of 10–11 meters is comparable with that of several other pterosaurs of that time. Its calculated mass of 200–250 kilograms exceeds that of the largest fossil Cenozoic bird, *Argentavis magnificens* from the latest Miocene (6 million years before present) of Argentina, which weighed 70–80 kilograms and had a wingspan of 6.5 meters. Among living flying birds, the South American condor *Vultur gryphus* has the longest wingspan (3.5 meters), and at 16 kilograms has the second highest mass behind the bustard *Ardeotis tardi* at 18 kilograms.[25]

These cases, taken from the history of land animals with a variety of habits, show that maximum size did not continue to rise beyond the Cretaceous period. One possible explanation is that endothermic mammals, with higher food requirements than ectotherms of equivalent size, emphasized high metabolic rates instead of extreme gigantism as the means to great power. As I noted earlier, Pleistocene elephants were still huge, and may have equaled or exceeded the absolute power of herbivorous dinosaurs. It would be interesting to know whether fully endothermic Cenozoic birds were able to achieve great power comparable with that of pterosaurs, which have also been considered endotherms but whose metabolic capacities are not well understood.

In the case of apex predators, another intriguing possibility is that cooperative hunting by smaller social animals could be an effective, alternative means of achieving great power without the drawbacks of maintaining the gargantuan sizes of solitary individuals. Social individuals and the groups to which they belong benefit from collaboration during hunts because they are better competitors with higher success rates. In its most sophisticated form, as seen in some carnivores, cooperative hunting requires coordination by individuals who are aware of each other as they chase or trap prey. Familiar examples include the spotted hyena (*Crocuta crocuta*) and lion (*Panthera leo*) of Africa and the gray wolf (*Canis lupus*) of North America and Eurasia. Less well known, but just as impressive, are tree-dwelling ants that build porous platforms or galleries out of leaves and fungi. Ant workers waiting below the platform snatch arthropods that land above them, pulling them through the perforations of the trap. In the case of the South American ant *Azteca brevis*, the largest prey captured in this collaborative way was 48.7 times the mass of individual workers. Among birds, pairs of hawks, falcons, eagles, and jaegers are more effective at capturing prey than are individuals hunting alone. Pairs of the Aplomado falcon (*Falco femoralis*) in eastern Mexico succeeded in capturing birds in 44% of attempts, whereas solitary individuals had only a 10% success rate. Some tropical spiders build communal webs, which catch larger prey than do webs of similar-sized solitary spiders.[26]

The importance of coordinated hunting is that it enables a group of relatively small predators to act as a single large entity. This is most obvious for social insects like ants, in which individual workers are analogous to the cells of individuals and in which the colony functions as a superorganism. A colony of the aggressive arboreal weaver ant *Oecophylla smaragdina* in northern Australia, which uses leaves to build communal nests, consisted of 376,635 ants collectively weighing 2.98 kilograms, about twice the mass of the heaviest solitary insect. In the modern fauna, group-hunting carnivores do not reach the sizes of solitary hunters such as the tiger. Humans, too, are relatively small group hunters when compared with the large solitary predators.[27]

The merits of coordinated group activity as an alternative to individual gigantism under highly competitive conditions are amply demonstrated by the status of these animals as top consumers on very large continents with long histories of intense competition. The carnivores in question are from the continents of Africa and Eurasia; and humans—the most powerful consumers of all—originated in Africa and modernized there and in Asia. Edward O. Wilson has repeatedly pointed out that some 80% of insect biomass comprises social insects, which are especially prominent as consumers in tropical forests. It is interesting that the mammal faunas of the island-like continents of Australia, Madagascar, and South America appear to have always lacked group-hunting species. This changed in South America when group-hunting dogs invaded from North America during the Pliocene.[28]

If the hypothesis that gigantic solitary predators were eventually replaced by smaller animals that hunted cooperatively is correct, the massive predatory dinosaurs and crocodiles of the Mesozoic era should have been solitary hunters. Evidence from rich bonebeds comprising many individuals belonging to single predatory dinosaur species points to the possibility that these animals were gregarious and perhaps social, but social behavior does not necessarily mean that individuals coordinated to hunt in packs. Likewise, fossil saber-toothed cats in North America could have been social, but whether they hunted in prides as lions do is doubtful. These cats are thought to have killed their prey with saber-like or scimitar-like upper canine teeth, a method that would not seem to lend itself to requiring help from others. At least for carnivores, the consensus is that pack-hunting emerged among wolves and other dogs during the Pliocene, primarily in open landscapes where long chases are possible.[29]

Large herbivores on land would gain nothing from cooperative foraging even when they move in herds as social animals. The fact that the largest Cenozoic plant-eaters are smaller than the largest Mesozoic ones therefore cannot be ascribed to the benefits of cooperative feeding. Growing food, as humans and many social insects do, does seem to require close collaboration among individuals. Gardening by insects is mostly a Cenozoic phenomenon but may go back to the mid-Cretaceous, whereas humans are the first relatively large animals to practice agriculture, beginning some ten thousand years ago.[30]

Post-Permian Marine Giants

In contrast to land animals, the largest marine plants and animals that have ever lived are still with us today or became extinct in very recent times. Among ocean-going plankton-feeders, the largest species of the Mesozoic was a pachycormid fish, *Leedsichthys problematicus* of the Middle Jurassic, with an estimated length of 16 meters. Later pachycormids were smaller, and there is a very long gap between the extinction of this group at the end of the Cretaceous and the large plankton-feeders of the Cenozoic era. These began as relatively small mobulid rays and cetotheriid mysticete whales of the Late Eocene and Late Oligocene respectively. A large Late Oligocene cetotheriid from Japan had a length of 8 meters. Gigantic balaeonopterid mysticetes, the group to which the blue whale (*Balaeonoptera musculus*) belongs, evolved during the Late Miocene, when a rorqual from the Pisco Formation of Peru already reached a length of 15.8 meters. Size repeatedly increased in several lineages thereafter. A fossil blue whale recovered from Pleistocene deposits in Italy (1.37 million years before present) was 26 meters long, comparable with the 33.5-meter length of living blue whales. This species is estimated to attain a mass of 170 tons in living examples. The next largest plankton-feeder giant living today is the whale shark (*Rhincodon typus*), which reaches a length of 21.3 meters and a mass of 34 tons.[31]

Apex predators in the pelagic realm are also larger today than at any time in Earth's history. The largest of these is the sperm whale (*Physeter macrocephalus*), a suction-feeder preying on large squid. It is said to attain a length of 28.5 meters, considerably larger than the great white shark (*Carcharodon carcharias*), with a mass of up to 3.32 tons and a maximum length of 7.1 meters. There were several extremely large fossil apex predators during the last 15 million years, but none quite reached the dimensions of the sperm whale. Prominent among these are the megatooth shark (*Otodus megalodon*), which became extinct after three million years before present during the Pliocene. This predator reached an estimated length of 15–16 meters. During the Middle Miocene, 12–13 million years before present, the huge raptorial sperm whale *Livyatan melvillei* from Peru has a three-meter-long head, teeth 36 centimeters in length, and a body length estimated at 13.5–17.5 meters. Its discoverers suggested that this animal tore large pieces of flesh from contemporary large whales much as the killer whale (*Orcinus orca*) does today. Earlier still, in the Late Eocene, the largest predators were basilosaurid whales such as *Basilosaurus isis*, which reached a length of ten meters. The apex predators of the Jurassic and first half of the Cretaceous were pliosaurid plesiosaurs, marine reptiles with maximum lengths of 10–11 meters. An exceptional Late Triassic ichthyosaur reptile, *Shastasaurus sikanniensis* from British Columbia, was perhaps 15 to as much as 20 meters long. The Middle Triassic ichthyosaur *Himalaysaurus*, living some 231 million years ago, is estimated to have had a body nine meters long, a little larger than the Early Triassic *Thalattoarcon sarcophagus*, the earliest large pelagic apex predator following the end-Permian mass extinction. With the exception of the ten-meter-long Late Permian fish *Helicoprion*, all Paleozoic pelagic apex predators were smaller than their largest Mesozoic and Cenozoic counterparts.[32]

Active bottom-feeding herbivores are latecomers to the oceans. Besides some doubtful Triassic fishes and placodonts, and a possible Paleocene pycnodont fish, confirmed herbivores date from the Early Eocene, about 50 million years before present. Gigantic herbivores—all sea cows of the order Sirenia or the extinct order Demostylia, and all weighing 5–10 tons—have existed in the ocean since at least the Oligocene. The largest of these, the Steller's sea cow (*Hydrodamalis gigas*), a kelp-eating species from the North Pacific, became extinct only three centuries ago.[33]

A possible explanation for the continued upward trend in maximum size of apex predators and herbivores in the sea is that cooperative hunting is rare among marine animals. The best-known pack hunter in the ocean is the killer whale, which attacks marine mammals in coordinated groups. This very large predator, up to eight meters long, was perhaps the only predator of large whales following the megatooth shark's extinction. It evolved during the Pleistocene, contemporaneously with the evolution of the largest mysticetes. The rise of this exceptional predator marks the beginning of group hunting among large marine predators. Group-hunting, technologically sophisticated humans have, of course, decimated whale populations. The social revolution that began on land with insects in the mid-Cretaceous and continued with group-hunting mammals during the last five million years of the Cenozoic thus came late to the oceans.[34]

An interesting question arises from the ever-increasing maximum sizes of pelagic plankton-feeders over the course of time. Does this pattern mean that the ocean's productivity has risen over time, or does it imply that the larger, more recent animals at the high end of the size spectrum sample more food from a larger area by virtue of their higher locomotor capacity and greater power? Perhaps both factors are at work. As an enabling factor, productivity must set limits on, and provide opportunities for, large size; but high productivity by itself does not guarantee that there are large-bodied species to take advantage of it. Endothermy enables a large swimming animal to cover a lot of territory and to locate particularly rich patches of plankton. It also makes the animal produce abundant feces and urine, which can speed nutrient cycling and thus stimulate production in the pelagic zone. There is, of course, a third possibility, which depends on the other two factors. Large active predators such as pliosaurs, sharks, and toothed whales act as potent agents of selection, favoring large size and rapid growth in their prey. I suspect that all three of these factors—a combination of enabling factors and the feedbacks among them—create circumstances in which gigantism is both possible and advantageous. With the arrival of industrial whaling by humans, however, this permissive yet competitive environment was replaced by a regime in which large size among whales was a liability. Not only did whalers prefer to kill large whales, but they acquired the technology to take them, a technology that no other predator or competitor in the ocean possessed.

Similar arguments can be made for patterns of gigantism on land. There had to be sufficient plant productivity to sustain viable populations of giant sauropods even if the metabolic requirements of these massive animals were lower than those of later and smaller mammals. As I shall discuss in chapter 7, there is good evidence that plant productivity did rise, especially from the second half of the Cretaceous onward, and that endothermic herbivores and apex predators enhanced this production by releasing copious fertilizer.

The emergence of overwhelming force in the form of a technologically formidable agent—the human species—demonstrates that, although large size will often be favored in competitive interactions, the extent to which maximum size can increase does not remain the same over time. Very large body size among active animals is no longer the advantage it once was, even though large size still comes with reproductive and competitive benefits despite the evolution of social hunters and the human juggernaut. This is so because smaller enemies are still important selective agents to this day.

Realms of the Less Powerful

Compared with the astounding sizes attained by herbivores, bulk suspension-feeders in the pelagic zone, and apex predators, the size spectrum of animals with lower metabolic rates and therefore less power is much more truncated. Within this more confined size range, however, enabling and selective factors still operate to produce some very large animals. These less active creatures are usually not targeted as prey by apex predators, but instead must contend with other less active enemies. In the case of benthic shell-breaking and shell-entering predators on shores and reefs, for example, the principal perpetrators are gastropods, octopuses, crustaceans, sea stars, fishes, birds, and small marine mammals. Except for the birds and mammals, these predators are ectotherms. In this world of armored prey and their predators, a pattern of increase in maximum size similar to that of more active marine giants is discernible.

Cenozoic herbivorous gastropods, for example, achieve larger sizes than their earlier counterparts. The largest herbivorous gastropods are the elongate Middle Eocene *Campanile parisiensis* from the Paris Basin of France, reaching a length of about one meter, and the modern, much bulkier Brazilian *Titanostrombus goliath*, with a shell 40 centimeters in maximum dimension. Mesozoic herbivores were much smaller, and no Paleozoic herbivorous gastropods are yet confirmed. Predatory gastropods reach a maximum length of 722 millimeters in the northern Australian and southeast Indonesian worm-eating *Syrinx aruanus*. Almost as large is the similarly spindle-shaped Floridian *Triplofusus papillosus*, a mollusc-feeder with a length of up to 600 millimeters. The globose gastropod-feeding northeast Australian volute *Melo amphora* (468 millimeters in largest dimension) and the Floridian echinoderm-feeding *Cassis madagasariensis* (400 millimeters) are also bulky predatory gastropods. Cretaceous predatory snails are all much smaller by comparison, with the largest species having a shell length of about 250 millimeters. There are no known pre-Cretaceous predatory gastropods.[35]

More active predatory sea stars and crustaceans are bigger. *Pycnopodia helianthoides*, the largest known sea star, reaches a diameter of

1.5 meters in the North Pacific. The stomatopod *Odontodactylus scyllarus* and the spiny lobster *Jasus edwardsi* both reach a length of 60 centimeters. Larger still are shallow-water octopuses such as the northeastern Pacific *Enteroctopus dofleini*, with an arm spread of 9.6 meters and a mass of 330 kilograms. The tropical to warm-temperate loggerhead turtle (*Caretta caretta*), a specialist on hard-shelled prey, attains a diameter of 110 centimeters.[36]

Maximum Size and Metabolism

The comparison between active vertebrate giants, with which most of this chapter has been concerned, and the slower, more armored giants of the shore and reef, reveals a striking pattern: high metabolic rates and activity levels offer the potential for much larger maximum sizes than do slow metabolic rates and limited mobility. Maximum size and maximum power are thus strongly correlated. Only when animals act in organized groups is this rule violated, for in these cases the group acts as one large entity. An increase in maximum size should therefore reflect an increase in maximum power.

The metabolic rates scaled to body size are only one-fifth to one-tenth as high in armored animals as in more active ectothermic fishes. The maximum sizes of these animals should therefore reflect local to regional productivity and other enabling factors, such as temperature, because individuals with less activity cannot sample as much area or volume of habitat for food as can active consumers. Active predators with high metabolic rates achieve larger maximum sizes and integrate their environments' enabling factors over larger scales of space and time. Their activity, in other words, makes them somewhat less dependent on regional limits than is the case for less powerful animals with a smaller reach. Productivity and temperature still matter, but their limitations on maximum size diminish as levels of activity increase. On the other hand, the size of the habitat—the area of a continent, the volume of an ocean basin or lake—becomes increasingly important as a determinant of maximum size, because very large animals cannot maintain viable populations in small environments.[37]

All else being equal, the evolution of activity-enhancing innovations such as the eukaryotic cell, locomotion, endothermy, and human weaponry enables organisms to achieve increasingly large sizes. In the next chapter, I turn my attention to one of these indicators of activity, namely, locomotion.

5

The Evolution of Motion

Nothing expresses power quite like the use of force. Animals bite, kick, stab, chew, swim, fly, shock, grasp, cut, squeeze, pound, and inject venom; they run, build, and display. Competitors counter these activities by doing the same things, as well as by evolving armor, toxins, threats, crypsis, and very large or very small size. Wielding and resisting power is expensive but essential. The larger the power budget, the more resources can be allocated to these functions. If competition favors more power and larger budgets, living things must invest more heavily in both offensive and defensive measures to the extent allowed by a permissive, productive environment. In short, escalation enforced by powerful enemies is to be expected along all the dimensions of power.

Animals and plants exist in a world of perpetual danger. To minimize risks to life and limb, organisms face several complementary choices. First, they can move away from danger, and the quicker the better. Second, they can hasten the time it takes to confront and defeat an enemy. Both options involve minimizing the time of an interaction. As I showed in chapter 1, time is the most important dimension of power, so minimizing time is a critical way to achieve high performance. For the same reason, maximizing the time required for an adversary to catch or subdue its victim benefits victims, because the longer the time an adversary spends, the greater is the risk that it will be interrupted or fall victim to another enemy.

To what extent is this expectation realized? Do the traits favored during selection change as power budgets increase? Did passivity give way to more active responses to selection?

Intuition might suggest that, as power budgets expand over time, organisms increasingly evolve adaptations that depend on a high throughput of energy. Instead of slowly accumulating a store of in-house energy for building a passively defended body characterized by permanent armor, toxicity, or distasteful deterrents, as might be expected in an organism with a small power budget, a high-powered organism would be better served by deploying force, on-demand chemical defense using readily synthesized compounds, or rapid movement to compete and defend itself. Greater activity like this requires access to outside sources of energy and rapid conversion to work by muscles and the production of fast-acting chemicals. How organisms respond adaptively to enemies therefore depends on the size of their power budgets.

This chapter and the next one concern the application of force. I first consider locomotion, the movement of part or all of an organism's body. This is arguably the first and most consequential use of mechanical force by living things, most obviously by animals but even by plants and microbes. For many organisms, locomotion is the only means for locating food, finding mates, and avoiding mobile enemies. It therefore provides far-reaching competitive advantages over a sedentary lifestyle, particularly when coupled with the ability to sense and interpret signals emanating from foes, allies, and hazardous environments.

One of the primary advantages of locomotion is that it enables organisms to sample a much larger part of the habitat for food or mates than is possible for organisms that are tethered in place as adults. Locomotion pays as long as the cost of searching does not exceed the rewards of large-scale reconnaissance. For predators that take prey singly, as well as for species in which mates are widely scattered, active search through locomotion will almost always provide a competitive advantage. Exceptions to such advantages occur in organisms that have found ways to draw food or mates to them, but this too requires a substantial investment in time and energy. Mating displays, including attractive flowers, are power-intensive; and sedentary suspension-feeders that draw food-laden currents to their filtering devices rely on ciliary movements of pumping. One way or another, movement of part or all of the body is an essential component of making a living.

The effects of locomotion on the environment itself are less well known but just as important. When moving through a medium like water or sediment, an animal body entrains particles with it, potentially transporting gases and nutrients from place to place and mixing the medium and its contents. Vertical migration up and down in the water column between night and day is an adaptive response of many planktonic animals and diatoms, transporting nutrient-rich bottom waters to the surface where phytoplankton can use them to photosynthesize and to increase the availability of food for the pelagic ecosystem. Nick Butterfield has further cogently argued that locomotion in water by early animals helped spread oxygen by transporting it from sites of photosynthesis to habitats where oxygen had been absent. Animals burrowing into sand and mud likewise not only redistribute nutrients by mixing the sediment in a process known as bioturbation, but also introduce oxygen into previously anoxic sediment layers. As a result, the added oxygen and nutrient mixing stimulate bacterial growth, which in turn make the infaunal environment beneath the sediment surface more congenial as a place to hide and feed. There is thus a positive feedback between locomotion and oxygenation: Mobility requires at least some oxygen, but it also helps expand the conditions where oxygen is available for animal activity.[1]

Locomotor performance can be measured in many ways depending on which aspect of movement is most effective under the circumstances. Acceleration and maneuverability are key for escaping or evading predators; their absolute performance levels depend on the locomotor abilities of the predators. Sit-and-wait (or ambush) predators also depend on rapid acceleration either of the body as a whole or of part of the body, such as the neck and head, a crab's claw, or a predatory snail's proboscis. High speed is useful in open habitats for pursuing prey or rushing away from the predator.

A problem both parties face is that high performance in locomotion is accompanied by wakes or other disturbances in the surrounding medium. These disruptions can be detected either by the enemy or the victim. For rapid movements, therefore, there is a premium on noise reduction, achieved by streamlining the body so that there is as little disturbance of the fluid particles flowing past the body as possible. This problem is especially acute in viscous media such as water, mud, or sand. A predatory copepod swimming at five centimeters per second in freshwater can detect smaller zooplankton swimming nearby, but is unable to detect passively sinking cladoceran crustaceans such as *Bosmina*, which cannot simply escape from the copepod by swimming. Even in air, locomotor silence can be very effective. Moths and butterflies, for example, must often evade birds and bats. Although they depend on good vision to identify and escape from flying enemies, it is notable that their slow wing movements are inaudible, at least to me. Quiet stepping while walking or running on the ground preserves an element of surprise for a stalking predator, as human hunters well know. Malaria mosquitoes (genus *Anopheles*) must take off with a heavy load of blood from their mammalian victims without being heard.[2]

Still another facet of locomotor performance is distance traveled. Long-distance endurance is critical to pursuers, but also to animals feeding on widely dispersed food sources, such as vultures seeking carrion and pollinators that are faithful to the flowers of rare plants. Long-distance migration as practiced by many herbivorous mammals, temperate birds, and large sharks and cetaceans enables animals to take advantage of foods whose distribution varies over the course of a year.

It has proven very difficult to estimate performance values of locomotion for fossil animals. Aside from a few estimates of walking or running speed in dinosaurs and cruising speeds in some fossil marine reptiles, the next best approach is to estimate times of origin of animal groups whose members include species with high-performance locomotor abilities. Comparisons of fossil animals with living counterparts with respect to limb proportions can yield inferences about types of locomotion in the past, but limb morphology and proportions turn out to be unreliable and often misleading. Nevertheless, the available indirect evidence indicates a general upward trend in locomotor performance in every dimension and in every medium.

Early Locomotion

Prokaryotes were likely the first organisms with the capacity to move. Together with the ability to detect magnetic fields and chemical gradients, these microbes evolved multiple ways of moving through water and on surfaces in their tiny viscous worlds. Movements were slow enough, and the medium was viscous enough, that fluid particles displaced by the body would have been difficult for other organisms to detect. Locomotion must therefore have served as a means to locate nutrients.[3]

Animal locomotion dates from at least 565 million years before present, during the Ediacaran period. Evidence comes from small horizontal trails made on the surface of microbial mats and sometimes beneath them. Movements were slow and likely involved peristalsis, muscular waves traveling along the body from front to back. Limbs had not yet evolved in these animals, which are interpreted to be ancestral bilaterians. Although drilling predators had evolved by the latest Ediacaran, they attacked sedentary tube-dwellers like the mineralized *Cloudina*. It is therefore unlikely that escape behavior from pursuing predators was an important component of early animal locomotion.[4]

Arthropods had evolved lateral and anterior appendages for locomotion and feeding by as early as 530 million years before present during the later Early Cambrian period. Besides enabling movement on the seafloor, these limbs could also be adapted for excavating burrows in sediments and for swimming in the waters above. Most of the Cambrian burrows in sand and mud were evidently nearshore resting shelters from surface-dwelling predators. Although some movement of wormlike animals within sediments already existed during the later Early Cambrian, widespread movement beneath the surface and for feeding in this infaunal environment did not evolve until the Late Silurian, 110–120 million years later, at about the same time that the first vascular land plants originated.[5]

Expansion of animals into the water column, either as drifting plankton or as swimmers, might also have been prompted by bottom-dwelling predators and competitors during the Early Cambrian. The earliest swimmers were mainly bivalved arthropods, some radiodonts, and early small crustaceans. A recently documented but not yet named possible cephalopod with an elongate, external, chambered shell from the Early to Middle Cambrian boundary sediments in Newfoundland might already have been a swimmer, but definite cephalopods swimming by jet propulsion are known from the Late Cambrian. If the early cephalopod did swim, it likely did so close to the seafloor.[6]

Maximum swimming speeds likely rose with the appearance of armored fishlike jawless vertebrates during the Ordovician. Early cartilaginous fishes (Chondrichthyes) and Silurian acanthodian jaw-bearing flashes were probably the most active swimmers of their time. The armored apex predator *Dunkleosteus terrelli* from the Late Devonian of Ohio would have been the fastest animal up to that time, but there were large predatory spiny-rayed fishes and tarphyceratid cephalopods with coiled shells that were also active swimmers. Speeds are difficult to estimate for any of these animals, but it is clear that highly streamlined swimmers and burrowers did not exist during the Early and Middle Paleozoic era.[7]

Mesozoic and Cenozoic

Both burrowers and swimmers achieved much higher levels of performance after the Paleozoic. Deep burrowing, often associated with intense reworking of the sediment by deposit-feeders and infaunal predators, became widespread during the Mesozoic and especially the Cenozoic eras, with crustaceans, sea cucumbers (Holothuroidea), heart urchins (Echinoidea), and even polychaete worms taking leading roles. Sand-burrowing gastropods, active bivalves, sea urchins, crabs, fishes, and small squid often became highly streamlined during the Late Cenozoic. The most intense bioturbators today and for the last few million years are rays (Batoidea), gray whales (*Eschrichtius robustus*), and walruses (*Odobenus rosmarus*), all predators of bivalves and other infaunal animals. Gray whales today are said to rework 500,000 cubic centimeters of sediment per day when feeding in summer in the Bering and Chukchi Seas.[8]

The Mesozoic also witnessed the evolution of the first truly fast large swimmers, animals with a stiff streamlined body and a hydrofoil-like tail attached to the body by a thin peduncle, powered by large trunk muscles. The earliest of these were ichthyosaurs, which following their Early Triassic origins became fishlike open-water swimmers during the Late Triassic. Ryosuke Motani's calculations indicate that these animals could sustain cruising speeds of 2–3 meters per second, somewhat slower than the speeds of later counterparts. Other fast Mesozoic swimmers include some plesiosaurs and, during the Late Cretaceous, some seagoing mosasaur lizards. During the Cenozoic these reptiles were replaced by lamnid sharks, scombroid fishes including tunas, and cetaceans. Some of these pelagic newcomers can reach astonishing speeds. Dolphins (Delphinidae) can swim up to 15 meters per second, whereas burst speeds of the tunas *Acanthocybium solanderi* (length 113 centimeters) and *Thunnus albacares* (98 centimeters) can achieve more than 20 meters per second. All these fast post-Paleozoic swimmers except the barracuda *Sphyraena* (130 centimeters, 12.2 meters per second) are known or inferred to be endothermic. Many smaller ectothermic ray-finned bony fishes can also swim remarkably fast: the 80-centimeter freshwater predator *Esox lucius* has a burst speed of 3 meters per second; a 60-centimeter eel (*Anguilla anguilla*) can manage 1.4 meters per second; and the small 6.5-centimeter *Gobius minutus* swims maximally at 27 centimeters per second.[9]

Jet-propelled cephalopods comprise one group of animals in which a general upward trend in speed and maneuverability is particularly obvious. During the Paleozoic, most cephalopods had an external, chambered shell, which in early forms was straight or gently curved. Beginning in the Early Devonian, many lineages evolved coiled shells, which were lighter in weight. None of these Paleozoic forms could be described as streamlined. Even during the Mesozoic era, when some of the coiled ammonoid cephalopods became disc-shaped and relatively streamlined, the volume of water that could be ejected during each propulsion step was small, indicating that these open-water cephalopods likely would not have exceeded a speed of 50 centimeters per second. Shell loss, and therefore greater speed and maneuverability, began in some lineages as early as the Early Devonian, and evolved convergently in other lineages in the Mesozoic. Most of these, including the belemnites with a heavily internal cylindrical shell, must have been slow swimmers compared with the squids that predominated after the extinction of all but one lineage of shell-bearing cephalopods at the end of the Cretaceous. Modern squids can be very fast indeed. The jumbo or Humboldt squid *Dosidicus gigas* from California and Mexico, some 80 centimeters in length, can attain a burst speed of 30 meters per second.

Squids are far less constrained in speed than shell-bearing cephalopods because they can take in and expel a far greater volume of water and do not have to move a shell that, even when thin as in many ammonoids, still limits acceleration.[10]

The Land

The potential for high speed and maneuverability is even greater on land than in water owing to the much lower density and viscosity of air, which offers less resistance to movement. This liberating effect is most apparent in small animals. The Saharan silver ant *Cataglyphus bombycina*, with a length of 11 millimeters, can walk at a speed of 85.5 centimeters per second, equivalent to 108 body lengths per second. It achieves this extraordinary speed by placing only three legs on the ground at any one time. The European lizard *Lacerta viridis*, weighing just 4.6 grams, flees at a running velocity of 3.7 meters per second. Comparably sized bottom-dwelling animals in water would maximally crawl or walk at less than one-hundredth of these speeds. For larger animals, too, fast running or walking under water is not really feasible. If large animals do move on the bottom, they either glide on cilia, as in some sand-dwelling gastropods, or move slowly using walking legs, as in lobsters and crabs. Sea stars (Asteroidea) depend on tube feet to move, and many gastropods use muscular waves along the ventral foot. Snails that escape sea stars or other gastropods often resort to leaping, falling passively from rocks, or even swimming to get away. Even some bivalves have evolved the capacity to swim using jet propulsion by clapping the valves in quick successive steps. Swimming has also evolved in a number of bottom-dwelling crab lineages.[11]

The potential for high-performance walking and running on land was not realized fully until the Cenozoic, hundreds of millions of years after the first animals colonized the land during the latest Cambrian or Early Ordovician. Before the arrival of four-footed vertebrates (tetrapods) on land during the Early Carboniferous, not later than 340 million years before present, all terrestrial animals were relatively slow arthropods. These creatures—mites, spiders, scorpions, centipedes, millipedes, and insects—continued to be the primary ground-dwelling walkers through most of the Late Carboniferous as well. By 290 million years before present, during the Cisuralian stage of the Early Permian, the first possible fast runner, the bipedal reptile *Eudibamus cursoris*, had appeared. Most Permian tetrapods, however, moved slowly on limbs that either sprawled to the side of the body or had a facultatively erect posture with the limbs directly beneath the body.[12]

Ground-based tetrapods of the Mesozoic achieved higher maximum speeds but did not reach Cenozoic levels of performance. Estimates of running speed based on dinosaur limb mechanics, trackway measurements, and calculations of stride length and hip height indicate a maximum velocity of 17.8 meters per second for the small Late Jurassic theropod *Compsognathus* (weight 3 kilograms). Somewhat slower were the Late Triassic *Dilophosaurus* (430 kilograms, 10.5 meters per second), the Late Jurassic apex predator *Allosaurus* (1400 kilograms, 9.4 meters per second), the Early Cretaceous *Velociraptor* (20 kilograms, 10.8 meters per second), and the gigantic Late Cretaceous *Tyrannosaurus* (6000 kilograms, 8.0 meters per second). Tracks made by a relatively small Early Cretaceous carcharodontosaurid or spinosaurid in Spain indicate speeds up to 12.4 meters per second. These dinosaur speeds are all for bipedal forms, and are similar to those in modern flightless birds such as the ostrich (*Struthio camelus*, 65.3 kilograms, 15.4 meters per second). Bipedal dinosaurs were likely faster than their quadrupedal ancestors.[13]

Maximum speeds of large tetrapods increased markedly after the Cretaceous, especially during the Pliocene, 5–3 million years before present. Pursuit predators with Pliocene origins are the fastest ground-dwelling animals today. These include the African cheetah (*Acinonyx jubatus*, 55 kilograms), with a speed of 26 to perhaps 29 meters per second. Somewhat slower are greyhounds (*Canis familiaris*, 25 kilograms) at 19.5 meters per second, and the wolf (*C. lupus*, 70 kilograms) and African hunting dog (*Lycaon pictus*, 20 kilograms) at respectively 18 and 19 meters per second. Earlier Cenozoic predators in North America were slower still, with more arboreal habits and a sit-and-wait ambush method of attack. Such predators still prevail today in forested habitats where very fast locomotion is prohibited by the cluttered environment of trees and lianas.[14]

The mammalian prey of the large pursuit predators are also extremely fast, perhaps in response to selection imposed by their enemies. One example is the North American pronghorn (*Antilocapra americana*), which at a weight of 50 kilograms can run 27 meters per second. There is currently no fast predator of this species, but two extinct candidates, the hyena *Chasmaporthetes* and the cheetah-like cat *Miracinonyx*, could have been effective selective agents in the evolution of high speed in the pronghorn. Africa, which is still home to large, speedy predators, also retains some very fast herbivorous prey mammals, with several antelope species exceeding speeds of 20 meters per second and one (*Gazella subgutturosa*), with a mass of 30 kilograms, running 27 meters per second. The human species, with a sprint speed of 12 meters per second and many features specialized for endurance running, also evolved in Africa. Elsewhere, small mammals like hares (*Lepus europaeus*, 4 kilograms) can bound at a speed of 20 meters per second. All these speeds would have been unheard of during the Mesozoic, even for small nimble theropods.[15]

Curiously, all the pursuit predators are from the very large landmasses of Africa, Eurasia, and North America, with the latter two having been connected north of the Pacific Ocean for much of the Cenozoic era. No pursuit predators are known, either living or extinct, from the more island-like continents of South America, Australia, and Madagascar. This absence could either reflect less intense competition and selection for high speed or limits on the territorial (and therefore population) size of pursuit predators on these relatively small landmasses. It is therefore perhaps even more surprising that Mesozoic theropod dinosaurs and their prey, which inhabited huge continents in both the northern and southern hemispheres, were much slower than their Cenozoic mammalian counterparts. Full endothermy may have enabled both high speed and endurance in mammals, whereas the inertial homeothermy of very large dinosaurs would have been incompatible with such high locomotor performance.[16]

Even among predators on the large continents, speed and endurance are not the only criteria for predatory success nor the only solution for prey to adapt to their enemies. Ambush predators, such as many cats (Felidae), rely on stealth and short-term sprints, and their prey depend either on large size or aggression with weapons like horns to thwart these predators. Such predator-prey relations and adaptations have likely been the norm through much of the history of land animals, whereas pursuit and endurance are recent and dramatic exceptions.[17]

The Rapid Deployment of Parts

Throughout this discussion of animal movement, I have emphasized speed and maneuverability of the whole body. Many animals, however, achieve the advantages of high locomotor performance by restricting rapid motion to only the head, tentacles, or other extensions. For many stealthy predators that go after small, fast-reacting, elusive prey, slow movement of the body toward the prey limits the ability of the victim to detect its enemy until it is too late, at which point the predator can strike by rapidly extending its capture device. This is evidently what many ocean-going squids do. Once within striking distance, they deploy the tentacles (or arms) to ensnare the victim. The extraordinarily long neck of several Mesozoic marine reptiles including the Middle Triassic *Tanystropheus* and *Dinocephalosaurus* and Middle Jurassic plesiosaurs enabled these predators to approach and catch small prey undetected and without the energetic cost of moving the entire body. No modern marine vertebrate has a comparably long neck, which in *Dinocephalosaurus* exceeded the length of the trunk. Terrestrial sauropod dinosaurs also had very long necks, which allowed them to explore a wide area for plant food while the body remained relatively stationary. Many lineages of fishes during the last 100 million years have evolved the ability to extend one or both jaws forward to catch elusive prey or small animals that are well hidden. In an extreme case, the labrid wrasse *Epibulus insidiator* from the tropical Pacific can extend its elongate jaws beyond the body by 21.4% of body length. The muraenid moray eel *Muraena retifera* can even protrude its pharyngeal jaw, located in the throat, out of the mouth to grasp struggling prey or probing human fingers. Portunid swimming crabs can catch fish by rapidly deploying their long, sharply toothed claw-bearing first pair of walking legs. As a final example, cone snails (Conidae) can catch and immobilize fish either by first enveloping and then stinging the prey in the very long proboscis or by releasing a harpoon-like tooth carrying venom from its radula. In the case of the cone snail *Gastridium geographus,* which uses the first method, the extended proboscis is three times the length of the shell, which in adults can reach 100 millimeters. With the possible exception of the squids and cone snails, animals relying on these rapidly deployed parts are restricted to small food items that in the case of prey are often difficult to approach without being detected. All the animal groups in which fast-acting body parts do the ambush work on prey are known or inferred to have post-Paleozoic origins.[18]

Flight

Flight in air offers animals not only the fastest form of locomotion but also the potential to cover enormous distances. On average, it is six times faster than running on the ground by animals of the same mass. Top speeds achieved by vertebrate fliers are impressive: 31.1 meters per second for the common swift (*Apus apus*), and 44 meters per second for the Brazilian free-tailed bat (*Tadarida brasiliensis*) in Texas, the latter weighing only 12 grams and capable of traveling up to 160 kilometers per night. Raptors, aided by gravity, achieve a speed of 50 meters per second when swooping down to the ground to catch prey. The godwit *Limosa lapponica* can fly for 12,000 kilometers without stopping as it migrates from Alaska to New Zealand across the open Pacific Ocean. The pollinating hummingbird *Phaethornis superciliosus* can fly for 40 meters between flowers at a speed of 11 meters per second. Insects are slower, but many tropical pollinating bees can travel for kilometers as they visit widely separated flowers in rainforests.[19]

Although active wing movements characterize most of these fliers, passive soaring is an important, cost-saving method of traveling very long distances, enabled by thermal updrafts of air. Fliers must generate lift by flapping the wings while taking off, but once aloft they can rely partly or sometimes almost entirely on soaring, as is the case for albatrosses

(Diomedeidae), frigatebirds (Fregatidae), and condors. The South American condor (*Vultur gryphus*), one of the heaviest flying birds weighing up to 16 kilograms, flap their wings for only 1% of the time they are aloft, and can cover a distance of 162 kilometers without flapping for five hours. The great frigatebird *Fregata minor* can stay in the air for months at a time thanks to soaring over the ocean beneath and sometimes inside cumulus clouds.[20]

Passive gliding using membranous extensions of the forelimbs or the front part of the body has evolved in many vertebrates, the earliest of which is the Late Permian reptile *Coelurosauravus*. Mammals, lizards, snakes, and frogs have each independently given rise to gliders, all using trees as launchpads. Although these animals have substantial control over the direction and rate of descent of gliding, the lift-generating structures do not flap as the wings of birds and insects do. They therefore require little power to operate.[21]

The evolution of active (that is, powered) flight began on land with insects about 340 million years ago. Devonian precursors may already have been able to glide in controlled directions from trees much as many nonflying ants do today. As with other gliding and flying vertebrates, escape from predators was likely a primary selective agency favoring aerial locomotion. Vertebrates achieved flight first during the Late Triassic with the evolution of pterosaurs, and later with birds during the Late Jurassic and bats in the Early Eocene. It is highly likely that both of the latter two groups evolved flight more than once, perhaps two to three times in Dromaeosauridae, up to two times in birds, and three times in bats.[22]

To escape from predation in the water, several groups of squids and fishes have evolved the ability to fly in air. Ocean-going ommastrephid squids up to 135 centimeters in mantle length fly by jet propulsion while in the air for three seconds, traveling a distance of 26.4–33.5 meters at speeds of 8.8–11.2 meters per second. Air speeds of these squids is 3.3–3.4 times faster than in water. The origin of flight in squids cannot easily be pinpointed, but most squids date from after the Cretaceous. Flying fishes, on the other hand, are known from the Early Middle Triassic onward.[23]

Rolling

Perhaps the only form of locomotion that has not been evolutionarily well explored by animals is movement by rolling. Michael LaBarbera observed that no animal has fashioned wheels, and pointed out that, even if this could be accomplished through development, animals would need flat terrain in order to make wheels functionally feasible and useful. Simon Conway Morris considered that the bacterial flagellum—and its archaeal equivalent, the archaellum—is the closest approximation to a wheel, but this rotatory structure is not used to propel the cell over a surface as human wheeled transport does. On land, several ants are capable of curling into a ball or spiral to effectively roll away from danger. The habit is also known in some caterpillars, a few desert spiders, African and Asian pangolins, and the Italian ant *Myrmecina graminicola*. In the case of the ant, rolling is active, powered by the antennae and hind limbs, attaining a speed of 40 centimeters per second. Dung beetles form balls of animal dung, which they roll away for later consumption. In some cases, the beetles roll atop their dung balls.[24]

The only rolling animals recognized thus far in marine shore environments are chitons (polyplacophoran molluscs). The foot muscles in these molluscs can transform the flattened or vaulted eight-plated body and shell into a ball, in which the head and tail valves almost touch but do not interlock. The function of this rolling-up habit, which is nearly universal among chitons, has long been a mystery, but recent experimental work shows that it appears to function primarily as a means to roll away passively from dangers such as predatory sea stars.[25]

Locomotion, Agency, and Time

The ability to move provides animals with agency. Mobile animals thus have some degree of autonomy in choosing where to live and work. Greater speed reduces the time required to find food, shelter, and mates, and to escape from danger, while greater distance traveled enables animals to find and exploit widely scattered resources. Maximum speed and maximum distance traveled have increased over time in both aquatic and terrestrial environments, but have reached higher values on land (and especially in air) than in water. The highest locomotor performance has been achieved during the last 100 million years, beginning in the mid-Cretaceous, with animals of the Cenozoic era realizing the fullest potential in the various modes of locomotion.

Although I have emphasized maximum performance in this chapter, broad-scale studies have shown a general upward trend in the fraction of marine animal genera whose members are categorized as active instead of passive. The competitive benefits of movement are so great and so widely applicable that selection for greater speed and longer distances covered has affected many lineages regardless of absolute power. Animals with limited mobility persist, of course, as exemplified by many of my favorite molluscs as well as armored vertebrates, but in a world where all competitors and predators are slow, either there are alternatives to high locomotor performance or the advantages of movement simply operate under severe constraints.[26]

Speed is often incompatible with the use of strong force or the ability passively to resist force. This aspect, together with the history of violence and the adaptations to it, is considered in the next chapter.

6

The Evolution of Violence

It is one thing to locate food and escape from enemies, but quite another to engage the enemy—or the victim—directly. Some forms of killing animals or eating plants are relatively passive and do not involve the application of strong forces. Drilling—making a small hole through a prey shell by means of mechanical abrasion and some chemical dissolution—is a generally time-consuming and very common form of predation on shell-bearing animals that by itself is not power-intensive; neither is the superficial grazing of soft seaweeds from rock surfaces. These methods work well when life is slow-paced, but when competition for food is intense, faster and more forceful methods are called for, and prey that coexist with predators with more potent weapons must either build stronger fortresses or themselves become more aggressive and retaliatory in their defense. Reducing the time of subjugation is at a premium for many predators that must compete for prey with other predators and with thieves, which in the realm of animals we call kleptoparasites.

Victims, on the other hand, have an evolutionary incentive to increase the time it takes a predator to subdue its prey, because the more time-consuming the interaction, the greater is the possibility that the predator will be interrupted or that it will simply give up. Alternatively, both parties may evolutionarily try to outdo each other in aggression and violence, leading to the kind of destructive and expensive arms race that modern nation-states seem bent on pursuing. Which of these pathways of evolution is taken depends on the power budgets of the parties involved and the capacity of the ecosystem in which the interactions take place to sustain expensive measures of violence and resistance.

In this chapter I trace the broad outlines of the history of force and resistance to it. As is the case throughout this book, I emphasize maxima in the full realization that even the least powerful animals must apply some force to obtain their food. The point as always is that the most forceful agents have an outsized influence on the evolution and distribution of species with which they interact and the ecosystem in which they reside. As I shall show, it isn't just the absolute forces that have increased, but also the greater variety of ways in which force is applied and resisted. Both potency and the distance over which forces are applied have increased enormously over time, not just among predators but even among herbivores, potential mates, and plants.

Armor and the Evolution of Biting

With its roots among single-celled organisms, predation has a long history before the evolution of animals. Once multicellular animals capable of locomotion appeared, however, it did not take long for them to evolve into predators and for other organisms to develop defenses against them. In fact, animal defense and predation appear at about the same time, 548–550 million years ago, toward the end of the Ediacaran. The earliest known method of predation by animals is drilling, as demonstrated by small holes in the wall of *Cloudina*, one of the earliest animals to produce a mineralized external skeleton. The perpetrator remains unknown, but it was likely soft-bodied. Its existence, together with the fact that some of the holes in the tube walls are incomplete, strongly implies that the primary function of the external skeleton was defense against predators.

In the ensuing Cambrian period, mineralized skeletons evolved independently in dozens of animal, protistan, and algal lineages. Beginning about 530 million years ago, trilobites were among the first animals to coopt skeletal parts for breaking the skeletons of prey animals. They deployed devices at the bases of their limbs to crush external skeletons with a force comparable with that of living horseshoe crabs (*Limulus polyphemus*), which are known to attack small bivalves. Even at this early stage in the history of predation, attacks were not always successful, as indicated by healed injuries in trilobites and small brachiopods. Predators, in other words, were potent selective agents, placing a premium on skeletal defenses that resist attacks passively.[1]

An early type of skeletal defense to emerge was enrollment in trilobites. During the Early Cambrian, trilobites including agnostids could flex the body so that the head and tail came together in a spiral configuration, protecting the ventral side of the body as well as the limbs. Later forms of enrollment enabled the head and tail to interlock, so that any predator would have difficulty penetrating the trilobite by unrolling it. Enrollment of this type is rare in marine animals today, possibly because an animal in the enrolled state can neither feed nor respire. Other Cambrian trilobites evolved curved spines, which point outward to form a bulwark when the animal curled into a spiral or ball. Both these defensive types could be effective against predators that swallowed their prey whole or used forces to break the skeleton.[2]

That the early defenses against marine predators all involve external armor is not surprising given that most bottom-dwelling animals in general, and predators in particular, were slow, and that armor is inconsistent with rapid escape maneuvers. If some prey species did rely primarily on evasion or escape, they might be difficult to detect because the speed involved would be slow enough that the usual streamlined shapes or other shapes indicating a quick getaway would have been unnecessary. Even today, armored gastropods whose primary defense consists of escaping from relatively slow sea stars are not notably streamlined.[3]

The formation of a mineralized skeleton appears to require the presence of oxygen. At and before the time the first mineralized skeletons appeared, however, sufficient oxygen was likely present only locally in coastal shallow waters, and only bottom-dwelling organisms developed external coverings hardened by calcium carbonate or in some cases silica or calcium phosphate. In other words, external armor may have been accessible to organisms only locally rather than globally as it became later, when oxygen levels rose and its availability reached most of the productive biosphere.[4]

Little is known about identity or capacity of Early Paleozoic predators capable of breaching the defenses of externally armored marine animals. The high incidence of openly coiled shells among Ordovician and Silurian gastropods and cephalopods indicates that these shells were vulnerable to predators, because much of the shell was left unbuttressed and therefore susceptible to lethal breakage. Instead, prevailing shell architectures of these Early Paleozoic animals enabled modest withdrawal of the vulnerable head and foot into the shell, aided in many gastropods by the presence of an organic or mineralized operculum that closed the shell's opening when the soft parts were retracted. Early shells had not, however, evolved very deep retraction of the soft parts, nor had they developed narrow shell openings or reinforced apertural rims lined with toothlike thickenings, all attributes that became more common during the Mesozoic and especially the Cenozoic eras. Likewise, whole-shell resistance against crushing, promoted by a thick-walled, tightly coiled, and therefore well-buttressed shell architecture, was also uncommon during the Paleozoic.[5]

The evolution of jaw-bearing fishlike vertebrates during the Silurian and of jawed cephalopods during the Early Devonian marks the beginning of stronger predatory weapons and a sustained increase in the incidence of antipredatory defenses in prey. These defenses include spines in echinoderms, rigid articulation in the skeletal plates surrounding the calyx of sedentary crinoids, and the development of spines and the capacity to cement the shell to the surface among brachiopods. These Mid-Paleozoic innovations indicate substantial escalation between predators and prey on the seafloor. In the Devonian, many groups of fishes, including placoderms and marine lungfishes, had evolved rounded or flattened teeth or even teeth fused into dental plates suitable for shell crushing.[6]

Despite the evolution of high-spired (or turreted) gastropod shells during the Devonian and the Late Paleozoic, enabling the snails' soft parts to withdraw relatively far back into the shell, rates of predation-induced shell repair remained modest during the Paleozoic and the Triassic stage of the Mesozoic as compared with the later Mesozoic and Cenozoic. The Mesozoic era witnessed at least two phases of escalation between skeleton-bearing benthic animals and their predators. The first, encompassing the Late Triassic and Jurassic periods, was marked by the evolution of cementing bivalves, sea urchins with tightly articulating skeletons of interlocking plates, and crablike decapod crustaceans in which the soft, vulnerable abdomen came to be situated beneath the body or protected inside the external shell of molluscs rather than being extended behind the body. It was not until the Late Cretaceous, about 100 million years ago, that rates of shell repair rose to modern levels. Among the novel shell defenses that appeared during this second phase were a narrow gastropod shell aperture, often obstructed with ridges and folds; a long anteriorly or dorsally directed shell extension (the siphonal canal) housing the proboscis used in feeding and chemical detection; a ventrally thickened and gravitationally stable gastropod shell supporting a large expanded foot, suitable for both armor and relatively rapid movement; and crenulated inner valve margins and complex hinge dentition in bivalves, enabling the valves to remain aligned while they are opened and closed and preventing predators from easy forcible entry.

There were corresponding increases in predatory force. Shell-cracking lobsters, stomatopods, fishes, marine reptiles, and much later birds and mammals with powerful devices for feeding proliferated during the Mesozoic and Cenozoic, beginning in the Late Triassic. The earliest crabs capable of breaking shells, as indicated by a differentiation between a robust tooth-bearing right claw and a smaller, more slender left claw, appeared during the Cenomanian period of the early Late Cretaceous. Some sea stars and gastropods evolved forceful means to open bivalve shells and to enter gastropod shells by pushing aside the operculum. In the case of gastropods, forceful entry is often accompanied in Cenozoic lineages by the application of anesthetics or venoms.[7]

In the modern fauna, the most powerful marine shell-breakers are some tropical crabs, especially members of the Carpillidae, Eriphiidae, and Parthenopidae, and vertebrates. Adult gastropods of the murivid species *Drupa morum* in the Indo-Pacific require up to 5 kilonewtons (5 thousand newtons) to break, yet crabs and puffers (*Diodon hystrix*) are capable of crushing their shells. The loggerhead turtle (*Caretta caretta*), which preys on large gastropods and other animals with skeletons, is the most powerful shell-crushing predator measured so far. A 90-centimeter-wide individual has a bite force of 1,762 newtons, and extrapolation to a large 110-centimeter turtle produced a calculated bite force of 2,105 newtons. Fortunately for the human handler, all these animals bite slowly, unlike the much quicker bites of fish-eaters and of venomous animals.[8]

The most powerful bites of marine animals are not reserved for shell-crushers, but for apex predators. The Late Devonian placoderm *Dunkleosteus terrelli*, possibly the most powerful marine animal of the Paleozoic, is calculated to have produced a bite force of as much as 4.4 kilonewtons at the jaw tip and 5.3 kilonewtons further back in the jaw, enough force to have pierced the armor of its fish prey. Later apex predators were stronger. The Late Eocene seagoing whale *Basilosaurus isis* generated a bite force at its third upper premolar of 16.4 kilonewtons, the highest force so far calculated for any mammal. By comparison, the great white shark *Carcharodon carcharias* is estimated to produce a bite force of 18.26 kilonewtons, while the much larger Pliocene megatooth shark *Otodus megalodon* is extrapolated to have delivered a bite of 108.5–182.2 kilonewtons. Like many terrestrial predators, these sharks likely relied not only on jaw musculature to deliver such powerful bites, but also on muscles in the trunk that enabled the head to magnify the bite.[9]

Few estimates of bite force have been made for powerful predators on land, but the available data indicate that they do not match the almost unbelievably powerful bites of the largest marine predators. The strongest terrestrial arthropod by far is the omnivorous coconut crab (*Birgus latro*), an island-dwelling hermit crab without a shell from the Indian and Pacific Oceans. This animal weighs up to four kilograms and is estimated to produce a bite force in its larger left claw (cheliped) of 3.3 kilonewtons. The greatest bite forces among living freshwater tetrapods are found in the crocodiles *Crocodylus proposus* (a species also known to go to sea), with an average bite force of 8.98 kilonewtons and a maximum bite at its molarlike back teeth of 16.4 to perhaps 27.5 kilonewtons. The Cretaceous crocodile *Deinosuchus riograndensis* could generate about 10.3 kilonewtons for an animal 11 meters long, whereas the Late Miocene South American apex predator *Purussaurus brasiliensis*, a crocodile weighing 8.3 tons and reaching a length of 12.5 meters, is thought to have produced a bite force of 69 kilonewtons, certainly enough to kill large terrestrial vertebrates. Tyrannosaurus rex, likely the most powerful dinosaur of the Mesozoic, could bite into bone with a force ranging from 8.5 to 34.5 kilonewtons, aided by massive jaw muscles and a powerful neck. Jurassic to Early Cretaceous apex predators—dinosaurs in the clade Allosauroidea—were more lightly built and, although they also had a powerful neck, would not have had such a strong bite.[10]

Mammalian apex predators have much less powerful bites than their reptilian counterparts. The Pleistocene sabertooth cat *Smilodon fatalis* is estimated to have had a bite force of 1.1 kilonewtons, almost twice that of the living jaguar (*Panthera onca*), with a force of 682 newtons. The force at the back teeth (carnassials) of the jaguar is about 2 kilonewtons. The most powerful bite of a living mammal is that of the spotted hyena (*Crocuta crocuta*), which can deliver a force of 4.5 kilonewtons. Mammals compensate for their relatively weak bites by masticating (or grinding) food between occluding teeth, made possible by jaw joints with three directions of motion: up-and-down, fore-aft, and side-to-side. Several lineages of large predatory mammals have evolved the ability to crack and ingest bone, as in Late Miocene to modern hyaenids (in Africa, Eurasia, and extinct in North America), Miocene to Pliocene borophagine dogs (North American Canidae), and the giant Early Miocene weasel *Megalictis* (family Mustelidae). Hyaenids crack bones mainly with the premolars, whereas the fossil dogs used molars. All these animals have exceptionally thick enamel in the teeth, enabling the teeth to resist the powerful shear and compressive forces needed to crack bone.[11]

The advantages of preparing and dividing food in the mouth by mastication likely did not evolve first in predators but in herbivores. Some degree of back-and-forth movement of the lower jaw was perhaps first possible in herbivorous mammal-like reptiles as early as the Early Permian, as in *Suminia* and the even earlier *Edaphosaurus*. Full-blown chewing, including side-to-side motion of the lower jaw, evolved 85 million years ago in hadrosaurs, and most notably in later mammals, whose multicusped teeth helped distribute the grinding forces associated with feeding on plants over a wider area of each tooth. Together with a complexly ridged surface and the ability to grow individual teeth during life, grinding enabled many lineages of ungulates to become specialized grazers, that is, consumers of silicon-rich grass.[12]

Two exceptional instances of chewing have come to light in aquatic vertebrates. One was discovered in freshwater South American stingrays of the genus *Potamotrigon,* in which food is prepared by up-and-down as well as transverse motion of the jaws. These fishes consume tough insects as well as small molluscs. It is unknown when this unusual specialization evolved. The other case is in slender aquatic North American salamanders of the genus *Siren,* which consume algae and other plants as well as some insects. The food is ground in the mouth with specialized ridges and some patches of upper and lower teeth. *Siren* and the less specialized genus *Pseudobranchus* are thought on phylogenetic grounds to have diverged early from other salamanders, perhaps during the Triassic.[13]

The substantial benefits of chewing make it surprising that extensive mastication of plant food evolved so late. Except for the hadrosaurs, the gigantic Mesozoic dinosaurs had simple teeth with which they grasped and ripped vegetation, which was then swallowed together without oral preparation. In the cavernous digestive system, plant material was broken down with the aid of bacteria (as in most other herbivores) and sometimes aided by stomach stones (gastroliths) that mechanically broke up the food. The teeth of sauropods, it turns out, were highly expendable, being replaced in as little as 35 days. Unlike the more permanent adult teeth of mammals, dinosaur teeth were likely much less resistant to abrasion and would have been unable to cope with the stresses of grinding or shearing. Birds, which also do not chew their food, lack teeth altogether and rely on gizzard stones to break down plant food mechanically. Most herbivorous birds are poor fliers or have become flightless, perhaps because the equipment needed to process plant material is bulky and heavy, interfering with effective aerial movement. In this connection it is notable that, as Gerald Mayr has noted, herbivorous birds—especially flightless ones—do best in places where mammalian predators are poorly represented, as on modern oceanic islands and in the extraordinarily rich Early Eocene fauna of Messel in Germany. This may also account for the absence of flightless or herbivorous bats.[14]

Puzzles

Four puzzles concerning how vertebrates deal with hard or tough foods deserve attention. The first is the fact that, whereas mammals and some dinosaurs masticate plant matter by grinding it between broad-crowned teeth in the mouth, tetrapods that consume hard-shelled prey do not tend to chew these foods. If shell-bearing prey are broken down in the mouth, they are crushed between broad occluding teeth in jaws whose only degree of freedom is opening and closing around a tight hinge at the jaw joints. This is done, for example, by the North Pacific sea otter (*Enhydra lutris*), which can also use rocks to smash the shells of molluscs. Coastal marine bears of the genus *Kolponomos* from the Early Miocene of the North Pacific also crushed shells, and used sabertooth canines to pry attached prey such as mussels from rocks. Other oral shell-crushers include South American reptiles: the large Middle Miocene crocodile *Natusuchus pebasensis*, some smaller living caimans, and lizards of the genus Dracaena.[15]

Most tetrapods that target molluscs as prey either swallow small prey whole, as in many ducks (Anatidae), or extract the soft parts outside the body before swallowing. Crows (Corvidae) and gulls (Laridae) smash shell-bearing molluscs by dropping them on rocks from aloft; mongooses (Herpestidae) and thrushes (Turdidae) break the shells of land snails on anvils; and oystercatchers of the genus *Haematopus* use the beak to dislodge prey and pick out the flesh before swallowing the edible parts without ingesting the shell.[16]

Some fishes, especially wrasses of the family Labridae, swallow shell-bearing prey whole but then grind the shells in the throat with greatly thickened pharyngeal bones. We have found fragments of small but very heavily armored gastropods in stomachs of the large Indo-Pacific wrasse *Coris aygula*. Freshwater cichlids in the Rift Lakes of East Africa also employ a pharyngeal mill to break the shells of their molluscan prey.[17]

The second intriguing conundrum concerning vertebrate diets and jaw forces is the surprisingly recent evolution of tetrapods specialized to consume hard seeds and nuts. Numerous birds and mammals have diets consisting mainly or exclusively of these resistant foods. Well-known examples include finches (Fringillidae), jays, and other corvoid birds, rodents, and primates, among many others. None of these specialized groups extend back in time to the Cretaceous. The earliest seeds indicating damage by gnawing rodents are from the latest Eocene of southern England. Earlier rodents are thought to have concentrated on fleshy fruits. Hard seeds and nuts evolved during the Early Cenozoic, perhaps as early as the Paleocene as indicated by the fossil record of Fagaceae (oaks, beeches, chestnuts and their allies) and Juglandaceae (walnuts and hickories). In the case of the latter family, hard nuts evolved from winged fruits, which are still produced by early-diverging members of the Juglandaceae. In fact, the tendency toward the production of large hard fruits and seeds is a Cenozoic phenomenon; Cretaceous seeds are generally very small, on the order of 6 cubic millimeters in size.[18]

In order to consume these hard plant foods, animals must bite through or otherwise remove the outer wall. Peccaries (Tayassuidae) crush very hard palm nuts between thickly enameled molars in jaws with a tight articulation allowing just one degree of freedom, opening and closing the mouth. Finches, grackles, and other birds crush seeds and nuts in the powerful beak, but others like jays peck holes in nuts. Rodents use the incisors to break open nuts, whereas primates apply both incisors and canines and secondarily the molars. Still others resort to breaking nuts open with stones. In short, hard nuts and seeds are either crushed in the beak or mouth, or broken outside the body; grinding is generally not used. Perhaps the occluding jaw surfaces simply cannot resist the forces and damage involved in grinding hard nuts and other brittle objects. The ability of many corvoids, rodents, and the acorn woodpecker (*Melanerpes formicivorus*, family Picidae) to hoard nuts and seeds for later use contributed to dietary specialization on these foods and to the evolution of plants that depend on the hoarders to disperse their seeds. Hoarding entails a good memory and, in the case of jays at least, awareness of potential thieves, capacities that may not have developed in Mesozoic dinosaurs.[19]

The third conundrum also involves vertebrates that consume hard foods, in this case other armored vertebrates. Although many predatory vertebrates have or had sufficient bite forces to crush turtles and other large, armored prey in the mouth, most killed their prey outside the body. The jaguar is one of the few living mammals with a high proportion of large turtles in its diet. Observations and inferences in the wild show that it kills turtles either by piercing the carapace with its canines or, for very large turtles, separating the dorsal carapace from the ventral plastron. Attacks from above are also employed against dangerous, aggressive prey such as anteaters. Perhaps the predator's mouth is not large enough or cannot open wide enough to accommodate a victim as large as a turtle. In any case, it is striking that so few predators have become specialized on a diet of armored large prey. Even armored catfishes (Siluriformes), syngnathiform pipefishes and sea horses, and ostraciid trunkfishes are not sought after much by other predatory fishes. In at least some cases, as my colleague Peter Wainwright informs me, the passive armor of these fishes is augmented by spines or toxins, which impede swallowing or make it unpalatable.[20]

The fourth puzzle is the reduction in maximum bite force among apex predators from the Late Cretaceous to modern times. One simple but incomplete hypothesis is that Cenozoic apex predators on land are smaller than their Mesozoic counterparts. Part of that decline in size can be attributed to the rise of social hunting in predatory mammals, and perhaps to a greater reliance on speed as a primary prey defense. There are two other possibilities. From the perspective of the prey, social organization can be a highly effective defense by dispersing aggression and retaliation among many prey individuals, effectively making a socially cohesive group equivalent to a single larger and more forceful solitary individual. Group defense is well known in some ungulates, primates, and especially social insects. In fact, the evolution of coordinated defense was likely a primary trigger for the emergence of eusociality in insects. In this connection it is interesting that there has been no reduction in bite forces among marine benthic predators since the Cretaceous, and there is no known case of social defense among benthic prey. In short, instead of relying on brute force, predators living in environments where sociality is common may have had to resort to quicker and less forceful means of subduing and killing prey.[21]

Venom and Electricity

Struggling prey pose a huge challenge even to the most powerful predators. They are capable not only of injuring the predator, but also of getting away once the predator has caught them. Powerful predators have evolved many ways to kill aggressive prey quickly. Besides a killing bite delivered by many vertebrates to the neck, two methods—envenomation and the use of electric shocks—have been used by some predators with relatively large power budgets to disable their prey with extraordinary speed. Both methods may have begun as defensive measures, and this remains the primary function of stinging by bees, diadematid sea urchins, and *Acanthaster* (the crown-of-thorns sea star). Nonetheless, both envenomation and electrical shocking are primarily employed by predators. The venomous stings of bees and herbivorous ants evolved in predatory ancestors.[22]

Envenomation—the injection of toxins into the body of another animal—is both ancient and phylogenetically widespread. Assuming that early cnidarians (jellyfish, sea anemones, corals and their allies) already had stinging nematocysts on their tentacles, they were likely the first animals to use envenomation, perhaps as early as the Late Ediacaran. Centipedes and scorpions were the first venomous animals on land. Caecilian amphibians—burrowing legless animals recently demonstrated to be venomous—may have originated as early as the Early Permian, but the major venomous clade of reptiles, a large group of lizards and snakes in the order Squamata, likely had Late Triassic or Early Jurassic origins. A few venomous mammals are known beginning in the Paleocene shortly after the end-Cretaceous extinction. On the seafloor, the only venomous mobile predators are small octopuses, a few blennies (Blenniidae), and two groups of gastropods, the highly diverse Conoidea including the lethally venomous fish-eating and mollusc-eating members of the Conidae, and the Colubrariidae, which suck blood from fishes. These molluscs all appear to have post-Cretaceous origins.[23]

Plants, too, can inject venom, as anyone familiar with nettles (Urticaceae) can attest. Here the function is entirely defensive. Deer avoid nettles but insects do not. As far as I can tell, stinging hairs are confined to angiosperms despite the fact that glandular hairs have existed since at least the Carboniferous. There appear to be no stinging hairs in ferns or gymnosperms, nor are any present in grasses (Poaceae), orchids (Orchidaceae), or lilies and their allies (Liliales, Amaryllidales, and Asparagales). The earliest stinging plant recovered thus far is a nettle from the Early Eocene of British Columbia, implying that the injection of venom by plants is one of several defenses that particularly targets mammals. The horror of a venomous insectivorous plant has not yet materialized.[24]

It is curious to note that some large groups of predators have never evolved the ability to inject venom into struggling prey. Prominent among these are sea stars (Asteroidea), birds, turtles, the mammalian order Carnivora, and decapod crustaceans. Likewise, there is no persuasive evidence of envenomation by Mesozoic dinosaurs. The inability of birds and turtles to envenomate prey is perhaps a reflection of the absence of oral teeth in these groups, but in other cases it remains a mystery.

Catching and subduing prey by electric shocks is one of the most recent novel methods of applying force. Although electrical communication is fundamental to the nervous system of all animals, the use of electricity as defense and as a means of predation is unique to six lineages of fishes, each of which evolved the capacity independently. Well-known examples include electric rays (Torpedinidae), electric eels, and electric catfishes (Siluroidea). South American electric eels such as the socially hunting *Electrophorus electricus* can stun multiple fish at once with a wallop of up to 860 volts. Although the generation of electricity is metabolically expensive, it is highly effective in turbid waters—both fresh and salt—where other senses and methods are less effective. Torpedinids could have had this capacity as early as the mid-Cretaceous, but most electric fishes appear to have Cenozoic origins.[25]

Given that electric signals travel well in water, one might have expected the generation of electric currents to be widespread among animals living in water. But this is not the case. Only fish have evolved the capacity. Highly active cephalopods, which in many ways act like and converge upon fishes, are not electrogenic; and toothed whales, which can stun their fish prey with bursts of sound, likewise have not resorted to electric shocks to achieve the same result. We can all be grateful that there are no electric venomous animals: Apparently one or the other weapon system suffices.

Remote Violence

Given the extremely high diversity of animal weaponry, it is surprising that so few animals have evolved methods to attack competitors and prey remotely. Cobras squirt venom into the eyes of mammalian prey, and the cone snails mentioned above release a harpoon-like tooth laden with venom to kill fish. Skunks (*Mephitis* and related Mustelidae), millipedes, and many beetles release noxious fluids to ward off enemies, and many sea cucumbers (Holothuroidea) release threads that are incredibly sticky under water, presumably as a defense against would-be predators. Most of these cases involve short distances, and devices targeting prey at a distance work in conjunction with potent venoms.

As in so many other ways, humans stand out from other animals in having perfected weapons that can kill or maim at long range. The trend began with throwing, an ability found only in humans and to a lesser extent chimpanzees, for which the arms and shoulders are particularly well adapted. Arrows, catapults, harpoons, guns, bombs, and drones followed, each deadlier than the last, and all designed to kill adversaries and prey. Besides providing for a vastly greater reach—an important component of power—these weapons also enabled humans to kill animals much larger than individual hunters. This remarkable evolution of long-range weapons is due entirely to the cultural acquisition of materials outside the body, together with techniques for fashioning these materials, from bone and wood to metals and gunpowder, into formidable instruments of violence.[26]

Group warfare is another transgression that, although not quite unique to humans, sets us apart from most other species. War requires coordinated group action and is, as nation-states have discovered through the ages, an expensive—and by no means always successful—activity. Humans have likely always experienced high levels of personal violence, but the advent of territoriality made warfare possible and sometimes necessary either to protect home turf or to take territory from rivals. It is coordinated warfare or its prospects that propelled human societies to develop increasingly long-range weapons.

Conflict between organized groups occurs also in ants as well as in the banded mongoose (*Mungos mungo*). Although such warfare is extremely violent, accompanied in the mongoose by much shrieking and mayhem as in human battles, nonhuman warriors do not resort to weapons fashioned outside the body. Confrontations involve body-to-body combat combined in the mongoose with visual assessments of opposing lines of combatants. Given that warfare is a social phenomenon requiring and favoring cooperation, we may confidently infer that it is a geologically recent development representing a substantial escalation in the scale of violence.[27]

How Much Is for Show?

Maintaining and using weapons of force is expensive. For this reason, some researchers have suggested that weapons have often developed from devices initially intended to inflict injury or death into less destructive devices, which function as warnings or in ritualistic combat. I am skeptical of this claim. There is little doubt that some animal weapons have been partially coopted into display structures, mainly through the action of mate-related selection. Horns, antlers, head shields, teeth, and crab claws often function as both weapons for defense or for killing struggling prey as well as for sexual displays, but in no case is the role of injury-inducing aggression entirely lost or even compromised. In fact, a warning or deterrence function would quickly become ineffective if the capacity to inflict injury were reduced or eliminated. Weapons for show, in other words, work as long as the possibility for violence remains.

Warning of potential danger at a distance by means of conspicuous colors, sounds, size, or startling behavior is certainly useful in preventing confrontations and reducing the chance of injury, but in animals with weapons it is usually an honest signal of potential mayhem if contact is made. There are, of course, numerous examples in which animals without chemical or mechanical weapons mimic the warning signals of truly dangerous animals, but such mimicry depends critically on the animals being mimicked retaining their weapons.[28]

It is, in fact, possible to make the opposite case, that warnings and ritualistic fighting become more common—and more necessary—as the potential for lethal interaction rises. This is, in fact, the macabre basis for the concept of mutually assured destruction, the idea that nuclear weapons are so powerful that rational leaders in the United States and the Soviet Union during the Cold War of the twentieth century would not want to deploy them. Steven Pinker has argued that violence of all kinds among humans has decreased relative to increasing population size, particularly with the emergence of centralized policing and evolving social norms. This trend, however, took place against a backdrop of vastly increasing power and reach of weapons as well as of intense competition involving nonviolent means. Today much of this competition is economic and political, with threats of violence not far beneath the surface.[29]

Warning displays of danger are by no means confined to animals. The strikingly creative Israeli botanist Simka Lev-Yadun has cogently speculated that plant spines are often conspicuously different in color from the green parts of plants, and that they are therefore visible to and avoided by herbivorous mammals. The whistling acacia (*Vachellia drepanolobium*) in Africa, moreover, may create a threatening sound to would-be herbivores by virtue of ant-inhabited cavities in its thorns, which emit a whistling noise when the wind blows. Warnings are, of course, very common also when the defense is a toxin, as in many insects, frogs, and plants. As with weapons, these warnings must always be backed up by the real physical or chemical defense: They are not all for show.[30]

Land, Water, and the Diversification of Violence

If there is one overriding conclusion that emerges from the history of violence, it is that the pursuit of power has followed increasingly varied trajectories. Force production and passive resistance to it were dominant themes on land until the Late Mesozoic and remain so among benthic predators and their prey today. With greater mobility and the evolution of social organization, however, this reliance on exerting and resisting strong forces was augmented by quicker, more dispersed or remotely applied forms of defense and protection. These newer, more active methods are, if anything, even more lethal—and more powerful—than the old ones. They represent an increase in the number of ways in which power is wielded by one side and diminished by the other.

That these changes are more apparent on land than in aquatic habitats reflects the greater freedom allowed in the medium of air than in the denser, more viscous medium of water. To be sure, some highly active and powerful animals have evolved in the sea, but many of these are derived from a land-dwelling ancestor. This is particularly striking among tetrapods, at least 70 independent lineages of which have colonized marine habitats during the Mesozoic and Cenozoic eras. Their ability to breathe air while remaining active has eliminated some of the restrictions on respiration imposed by animals in water that rely on gills or on diffusion of gases across the outer integument. Most of the world's marine apex predators and large herbivores have terrestrial origins and helped transform nearly all marine environments adaptively to become more like ecosystems on land. This pervasive evolutionary influence of land life extends to many other aspects of ecological power too, as I shall discuss in the next chapter.[31]

7

The Power of Economies

The power of living things is never gained or lost in isolation. No matter how much power is due to individual agency, it depends critically on the activity of, and interactions with, other life- forms. The energy component of power comes from the environment, much of which is either alive or strongly influenced by the work of organisms. In short, each living thing depends on the economy (or ecosystem) in which it competes. An economy, moreover, also has power, and it is this power that provides the potential for its members to work and evolve.

Where does economy-wide power come from? Individuals contribute to it, of course, but much of the system's power comes from rates of collective primary production and the feedbacks with consumption that regulate production. Without the intricate interdependencies among life-forms with different functions and performance levels, there would be no economic system, no evolution, and indeed no life. Like the agency of life itself, an economy is an emergent entity with properties that its components do not possess. It is the interactions and interdependencies among members—competition, collaboration, and trade—that confer these emergent traits of economic systems. Just as individuals work because of the interdependencies among their parts, economies work because their semiautonomous members suppress some of their self-interest for the common good of the whole.

My contention in this chapter is that all economies share fundamental emergent properties and processes, powered by competition for resources, and that their power-related upward trajectories through history should be similar regardless of whether the economy comprises life-forms of the past or is dominated and largely constructed by modern humans. To be sure, economies vary in space and over time, and many have developed unique characteristics just as evolving lineages have done. Nevertheless, the unity of economic organization is striking and places the human economy in a new and broader context from which lessons or even predictions about the future might be drawn.

I shall also show that ecosystems from the beginning of life to the present have increased in the magnitude of all their emergent properties. Except for relatively brief reversals during and immediately after great crises, there has been a general upward trend in all economic emergent properties, from power to diversity, disparity, and the resistance to external disruptions. Productivity and efficiency have risen over time, as has the ability of living things to destabilize economies. Whether these trends can continue in the face of the extraordinary monopolistic power of the human economy is a question I shall consider in chapter 8.

Emergent Properties

All economies depend on production by plants and other organisms that convert external energy and materials into biomass. An economy cannot be viable without consumption of these primary producers, nor can it persist for long without decomposition and recycling, the conversion of waste products and dead organisms back into materials that the primary producers can use. Sustainable economic systems are therefore circular, where all the processes depend on all the others. The efficiency of the economy depends on how much matter and energy become unusable and are thus lost from the system or even from the biosphere as a whole. The more biomass and raw materials stay in the active economy, the more efficient the economy becomes and the fewer resources disappear out of reach.

The processes of production, consumption, and recycling are continuous, and therefore can be expressed as rates. The power of an economy thus derives from the rates at which these processes are carried out, which collectively can be summarized as the rate of material turnover.

Calculated rates are not always constant. The longer the time interval over which a rate is measured, the lower is the overall estimate of power, because that time interval incorporates episodes of slow turnover. In an ecosystem that operates in seasonal climates, for example, rates calculated on an annual basis will be lower than those calculated for the warm or wet season alone, when activity would be highest. In the geological record, where rates are difficult to infer, such nuances may matter little, and economic activity must be measured mainly by preservable proxies that in living organisms indicate or correlate with metabolic rates.

Besides these emergent continuous processes, there are static properties that describe an economy but not its individual units. These include diversity—the number of species, occupations, capacities, and adaptations, for example—as well as the distribution of power among components. As I noted in chapter 3, this is where the entire distribution of performance—wealth, size, income, speed, and force—becomes important, because the distribution of these aspects of power expresses the degree of disparity (or inequality) in the system.

Economies have additional emergent properties related to their sustainability. Disruptions are ubiquitous. Their magnitude and duration vary according to a power law, whereby most disruptions are minor but some, such as mass extinctions or deep depressions, are catastrophic. Economies must be able to absorb shocks to maintain stability so that the interdependencies and feedbacks are maintained. Moreover, if the system does collapse or suffer substantial losses, the degraded system in the aftermath of a disaster must be resilient enough to reestablish the relationships among its members that confer stability.

How economies respond to disruptions depends not only on the magnitude of the event but also on which parts of the economy are most vulnerable. If, for example, a disruption targets the most powerful consumers in the system (the demand side of the economy), there will be collateral repercussions to the survivors, mostly because the positive feedbacks that these powerful members maintained are weakened or entirely dissolved. Yet the effects would likely be less destructive than if the disruption interfered with primary production, on which all members of the system depend. Such disruptions thus compromise supply and eventually demand as well. Supply and demand are, of course, always linked, and never independent of each other, but the source of economic instability, and which part of the economy is affected first, determines how the economy as a whole responds. The distinction between triggering mechanisms brings into focus the question which attributes of an economy confer resistance to disruptions and which ones provide a cushion.

Given the prevalence of destabilizing agencies, we can ask how an economy's emergent properties—power, diversity, and disparity—relate to its stability and resilience. Does higher diversity, for example, offer protection through redundancy against disruptions? What might conclusions about the history of ecosystems say about the vulnerabilities of the modern human economy?

Primary Productivity

Given its role in supplying economies with organic nutrients, primary production is an appropriate process with which to begin an inquiry into the power-related history of ecosystems. When life was anaerobic during the Early Archean, before three billion years ago, primary production was carried out by bacteria using light to produce organic matter without the release of oxygen. Donald Canfield and his colleagues estimated that productivity at this time was at most 7% that of modern marine values. Hydrogen-based and iron-based metabolisms, the most common forms of production of the Early Archean, would have yielded less than 0.1% of modern marine productivity. This low level was also endorsed by other analysts, all based on isotopic values and on models with many assumptions.[1]

Photosynthesis that liberates oxygen is today the most important form of primary production, ultimately supporting all herbivores and most other forms of life, even in zones where light does not penetrate. Phylogenetic and biochemical evidence points to the origin of this form of photosynthesis and of the cyanobacteria in which it arose about three billion years ago. For a period of 360 to perhaps 500 million years thereafter, however, photosynthesis contributed only very locally to primary production and the release of oxygen, because most of the oxygen reacted with reduced compounds and therefore did not stay in free form for long. By the end of the Archean, about 2.5 billion years ago, marine productivity as inferred from the availability of phosphorus was still only about 7% of modern values. With the Great Oxidation Event, beginning about 2.45 billion years ago and continuing off and on to 2.25 billion years ago, productivity may have risen somewhat, but all primary production remained in the realm of prokaryotes until at least 720 million years before present.[2]

During the early phases of oxygenation of the biosphere, the accumulation of oxygen was prevented by the high abundance of compounds that are reduced, that is, stable in the absence of oxygen. As oxygen was liberated, these compounds were progressively oxidized, keeping oxygen at levels below 2% of modern values before the Great Oxidation Event. The long-term burial of organic matter, which is not oxidized, enabled levels of oxygen to rise and for animals using respiration—in effect, the reverse of photosynthesis—to evolve and thrive.[3]

Subsequent plant innovations would have led to substantial increases in the rate of primary production. Between 650 and 635 million years before present, during the Late Cryogenian period of the Neoproterozoic era, planktonic and perhaps benthic green algae began to replace prokaryotes as the predominant marine primary producers. During the Late Ediacaran period, some 85–100 million years later, the beginning of bioturbation (sediment mixing by organisms) made more habitats available for algae, and for oxygen-requiring animals. Microbial life on land would also have contributed to the world's primary productivity by this time.[4]

The first signs of possible land plants date from 480 million years before present, during the Tremadocian stage of the Early Ordovician period. The available record consists of spore types and clusters diagnostic of embryophytes, but no material of photosynthesizing tissue has yet been recovered in Ordovician rocks. Isotopic evidence, however, indicates a rise in oxygen at about 458 million years before present, likely the result of increasing coverage of the land by bryophytes (liverworts, mosses, and hornworts). It was not until about 430 million years ago, during the Middle Silurian, that vegetative tissues of *Cooksonia*, the earliest known vascular plant, appeared. These events are associated with small but significant increases in the rate of weathering of minerals from rocks on land to form the first thin soils. Experiments with living bryophytes imply that the earliest soils, formed by thalloid liverworts and hornworts, would have been about one centimeter thick, whereas somewhat later leafy mosses would have created soils with thicknesses on the order of seven centimeters. Moreover, soils became increasingly organic-rich and therefore capable of supporting more productive plants.[5]

Photosynthesis in the earliest vascular land plants occurred on upright stems, which were initially simple but in later Devonian plants became branched. A substantial increase in the rate of photosynthesis must have begun by the Middle Devonian, when the first plants with leaves arose. Recent work on the phylogeny and embryonic development of plants indicates that leaves—flattened structures with veins—originated as branches connected by tissue, and that they arose independently as many as nine times in various lineages of monilophytes (ferns and their allies), spermatophytes (seed plants), and lycophytes (club mosses and their allies). The evolution of leaves is associated with a sharp increase in the amount of carbon dioxide and an increase in oxygen in the atmosphere. Whether this atmospheric evolution represents the cause of leaf evolution or a consequence is unclear. Some have argued that leaves would have been impractical at a time when carbon dioxide levels were still high, because they would have had few stomates through which evaporation could have cooled the leaf when in the full sun. As carbon dioxide levels dropped, stomate density increased and large expanses of photosynthetic tissue could be exposed to the sun as more water was transpired. On the other hand, with a greater surface area for the reception of light and the transpiration of water, leaf-bearing plants would have gained a significant competitive advantage over leafless forms. As usual, therefore, causes and effects intertwine. In any case, the repeated evolution of veined leaves greatly increased terrestrial primary productivity.[6]

Not only did early land plants carry on photosynthesis on stems without leaves, but they also lacked roots. Rhizoid structures and a simple conducting system do characterize mosses, which are phylogenetically closest to vascular plants; but the earliest roots, still without root hairs and root caps, had evolved during the Early Devonian (Emsian epoch), about 407 million years ago, as represented in the Rhynie flora of Scotland. Roots with hairs and caps evolved later in lycophytes and independently in euphyllophytes (all other land plants). With their capacity to use nutrients from soil, aided by symbiotic fungi, roots greatly increased the resource base in land plants, and were therefore as instrumental as leaves in raising primary productivity.[7]

The evolution of a vascular system for transporting water and the products of photosynthesis opened up an entirely novel realm of adaptive possibilities, many of which were already realized by the Late Devonian. Notably these included the tree and vine (or liana) habits, both achieved by several independent lineages of plants. The forests these plants created were surely more productive than the low-growing vegetations that preceded them, and liberated sufficient oxygen for the Late Devonian atmosphere to resemble the modern version in terms of the abundance of free oxygen. Nevertheless, inspection of the leaves of forest trees—lycophytes, pteridosperms (seed ferns), calamitaleans (tree-like horsetails), and early gymnosperms—indicate markedly lower forest productivity. Their vein density, as measured by the length of veins in each square millimeter of leaf surface, was generally two millimeters, a value that in living ferns and gymnosperms is associated with a low photosynthetic capacity. Although the stems of lycophytes and medullosan pteridosperms have large conducting cells (tracheids), indicating rapid conduction of water, the parts of the stem containing them are relatively narrow, suggesting that the absolute rate of transpiration (and thus photosynthesis) was low. One has to wonder why productivity remained so low. One possibility is that herbivores were minor agents of selection. Although some herbivory is known back to the Middle Devonian, frequent leaf damage is unknown until the latest Carboniferous, when both insects and vertebrates evolved the herbivorous habit. The Middle Carboniferous development of leaves in which the veins are interconnected rather than openly branching is consistent with this scenario, because connectedness confers redundancy of transport within the leaf when a small herbivore should sever a vein. Open branching of leaf veins and a low premium on replacement of damaged photosynthetic tissues in plants with low productivity could therefore exist in a world with little or no herbivory.[8]

According to Kevin Boyce, Tim Brodribb, and their colleagues, few if any plants of the Paleozoic and the first 150 million years of the Mesozoic had the characteristics associated with high photosynthetic rates. This changed dramatically with the evolution of eudicotyledonous flowering plants (angiosperms), grasses, and angiosperms with C_4 photosynthesis, beginning in the mid-Cretaceous about 100 million years ago. These plants achieved high photosynthetic capacities by virtue of a high leaf-vein density, which rose from an average of 2 millimeters to a mean of 8 millimeters per square millimeter of leaf surface, with some species achieving 25. This increase was made possible by the evolution of smaller cells, enabling the distance between sites of photosynthesis and water- and nutrient-conducting veins to diminish. A consequence of this evolution was a faster water cycle, which in turn led to an increase in the concentration of carbon dioxide and nutrients in the soil, further enabling high productivity. As I shall discuss later, the evolution of herbivores with high metabolic rates and therefore a large appetite may well have spurred the rise in plant productivity, because plants would have responded to more intense consumption by producing cheaper, more productive foliage. Dead plant material, moreover, would have decayed more quickly and provided more raw material for re-use by plants.[9]

Two additional, geologically recent innovations in flowering plants brought primary productivity on land to its modern high level. One of these is nitrogen fixation, an energetically expensive conversion of inert nitrogen in the atmosphere to compounds such as nitrate and ammonium that plants can use. This process had evolved in bacteria by at least 2.5 billion years ago, before the Great Oxidation Event, consistent with the fact that the principal enzyme, nitrogenase, must function in the absence of oxygen. Something like 25 moles of ATP are needed to produce just one mole of nitrogen compounds. Nitrogen-fixing bacteria became symbiotic on the roots of several plant families, most notably the Fabaceae (beans, peas, and their allies), during the Late Cretaceous. The earliest fossil of this family, a pod from the Campanian epoch (80 million years before present) of Mexico, was the harbinger of a great post-Cretaceous diversification. As a result, primary productivity rose not only for members of the Fabaceae, but also for neighboring plants in tropical forests and in vegetations at higher latitudes, and enabled members of the Fabaceae to invest in nitrogen-rich toxic alkaloids, especially in their seeds. This development is a cogent reminder that bacteria in symbiosis with multicellular organisms continue to play crucial roles in ecosystem-level activities.[10]

The second important modernizing innovation is the grass habit, in which the growth zone of the blades lies close to or even below the ground, so that trampling animals and tip-grazing herbivores do not fatally compromise the ability of the plant to grow and photosynthesize. Although grasslike plants were already present during the Late Cretaceous and the first half of the Cenozoic, they became the dominant vegetation of drier regions after the Middle Miocene, about 15 million years before present. These new grasslands not only support a high diversity and abundance of herbivores, but they also compete for light and nutrients without making enormous investments in wood as trees do. It remains a fascinating puzzle why the grass habit evolved so late in the history of vegetation. Perhaps, as with high leaf-vein density, the evolution of hungry endothermic plant-eaters created the selective conditions favoring this novel adaptive type.[11]

Did marine plants—phytoplankton as well as seafloor algae and photosymbiotic animals—show increases in photosynthetic capacity in the way that land plants did? The evidence is often sketchy and circumstantial, but in general it appears to show that there was an overall upward trend in marine ecosystem-level primary productivity both in the pelagic and benthic realms, especially from the Cretaceous onward.

Early stages in the rise of marine productivity would have been the Cryogenian evolution of planktonic eukaryotic algae, adding a new category of primary producers to the already existing prokaryotic cyanobacteria. In addition, multiple lineages of algae evolved larger seaweeds during the Cambrian and Ordovician. During the latter period, branching algae evolved from simple tubular forms that characterized Cambrian ancestors. All these events are associated with, and perhaps were triggered by, increases in direct consumption of primary producers. The Ordovician also witnessed the evolution of the first photosymbiotic animals in the form of tabulate corals.[12]

The transition from phytoplankton dominated by green-algal types in the Paleozoic and Early Mesozoic to lineages with red-algal photosynthesizing plastids from the Jurassic onward represents another boost in productivity. These newer single-celled primary producers—coccolithophorids, dinoflagellates, diatoms, and foraminifers—have larger cells than their predecessors, and have a higher phosphorus content relative to carbon and nitrogen. This latter characteristic makes them both more nutritious to zooplankton and more productive. Their larger size, together with the protection they derive from their mineralized skeletons, makes these newer members of the pelagic ecosystem more resistant to consumption. Production rose still further with the evolution of rhizosolenid diatoms about 90 million years ago during the mid-Cretaceous.[13]

The most productive benthic plants in coastal marine habitats belong to four distinct categories, all of which evolved during or after the Cretaceous. On tropical reefs, the predominant photosynthesizers other than corals and encrusting coralline red algae are small, fast-growing turf-forming algae, which are heavily cropped by fishes. Like the fishes, these algae are primarily of Cenozoic age. Temperate rocky shores support larger, leafy brown, green, and red seaweeds whose times of origin apparently postdate the Jurassic. Also, latecomers to the sea are two categories of angiosperms: seagrasses belonging to three separate families, and mangrove and salt-marsh plants. Seagrasses, which are not true grasses, appear first in the Campanian epoch of the Late Cretaceous, about 80 million years before present. They form lush underwater meadows, which in the tropics are further enriched by calcareous brown and green algae. Besides drawing nutrients from the sediment—a capacity all but unknown in seaweeds—seagrasses owe part of their high productivity to facultatively symbiotic nitrogen-fixing bacteria on their roots. Their colonization of sandy and muddy habitats represents a major expansion of marine primary producers. Finally, mangrove trees on tropical shores and salt-marsh shrubs and herbs in the temperate and polar regions have transformed mudflats into productive ecosystems. Mangroves evolved during the latest Cretaceous, whereas salt-marsh plants, which belong to entirely different families, are much more recent, with origins no earlier than the Late Eocene, 35 million years before present. Some lycophytes, ferns, and conifers probably occupied similar tidal habitats during the Triassic, Jurassic, and Early Cretaceous periods, but they were likely less productive and not as widespread as the later angiosperm mangroves and salt-marsh plants.[14]

A compelling argument for the stimulatory effect of marine herbivores on marine plant production is provided by a study of North Pacific kelps and the recently extinct Steller's sea cow (*Hydrodamalis gigas*). This huge mammal fed predominantly in the algal canopy, thus enabling sunlight to reach not only the lower parts of the kelps themselves but also understory algae. The ravenous appetite of these animals would have produced large amounts of excrement, which fertilized the seaweeds. Moreover, because the seacows cropped the uppermost parts of the kelp, the surviving parts would have better resisted the strong flow associated with ocean waves, keeping losses of the seaweeds during winter to a minimum. If Oligocene to Late Miocene desmostylian mammals had similar habits, as seems likely, this feedback between grazing by endotherms and production by large brown seaweeds began at least 30 million years ago.[15]

Three conclusions stand out from this romp through the history of primary production. First, the ecosystem-wide increases in primary productivity, both on land and in the sea, are associated with evolutionary innovations, which often originated in multiple lineages. Second, the rise in productivity brought about increases in the standing biomass of primary producers. Finally, the spread of the innovations as well as the greater biomass are associated with increasingly intense herbivory. Without the production-enhancing innovations, there might not have been corresponding intensification of herbivory; without herbivory, however, the innovations might have arisen but not spread, because their primary benefit is to adapt to high rates of consumption. For photosynthesizers, primary productivity is power; for consumers, it is an enabling factor.

Unproductive ecosystems persist, of course, much as organisms with low power continue to thrive alongside more powerful ones. Nevertheless, it is likely that power has increased even in these unproductive habitats. Modern maritime tundra, nitrogen-poor peatlands, and coastal salt marshes and mangroves are dominated by angiosperms, whereas earlier versions of these habitats were vegetated by ferns and gymnosperms. Deserts, which are also occupied primarily by flowering plants, could represent relatively new ecosystems of low productivity.

Animal Metabolism

The mutual dependence between primary production and consumption implies that, if productivity rose over time, collective consumption should follow a similar trajectory. In chapters 4, 5, and 6, I already discussed the upward trends in maximum body size, locomotor performance, and the use of force in consumption and defense, all of which are aspects of power. There is, in addition, abundant evidence for rising metabolic rates among consumers as well as for increasing activity of animals on land and in water.[16]

The trend toward greater activity may be most apparent among mobile animals, but it is also evident among the ranks of animals with sedentary habits. It has long been known, for example, that barnacles (Cirripedia) and bivalved molluscs have displaced bivalved brachiopods and sessile echinoderms as the most abundant bottom-dwelling suspension-feeding animals in many marine habitats. During much of the Paleozoic, brachiopods were the most common bivalved animals on muddy and sandy bottoms as well as on reefs. Although they suffered enormous losses during the end-Permian crisis, they rebounded during the Triassic and Jurassic, declining thereafter in most shallow-water marine habitats. Today, brachiopods are common in cold, deep, nutrient-poor, and sediment-free habitats. Elsewhere, however, they are virtually absent. Most beachgoers won't find their shells washed up, and on reefs brachiopods are restricted to deep, shady crevices. The metabolic rates of these animals are much lower than those of most bivalved molluscs.

Whether Paleozoic brachiopods also had lower rates of growth, reproduction, respiration, and feeding than contemporaneous bivalves remains unclear, but the fact that most of the volume inside the shell is filled with mineralized structures that support the feeding organs indicates that their activity levels were indeed lower than those of bivalves, which lack these internal skeletal features. Some Late Paleozoic photosymbiotic brachiopods must have had higher metabolic rates, as most animals with symbiotic primary producers in their tissues do today relative to animals without symbionts, but the largest of these (species of *Gigantoproductus* in the Carboniferous, reaching a shell width of about 30 centimeters) is still small compared with some Late Cretaceous rudists (Hippuritoidea) and giant clams (Tridacnidae). Among bivalves, moreover, lineages with powerful ciliary currents generated in specialized gills (ctenidia) for filter feeding, and with separate siphons for taking in and expelling water, have proliferated at the expense of lineages with weaker currents and simpler ctenidia and in which siphons are typically absent.[17]

Metabolic rates in fossil animals cannot be measured directly, but they might correlate with shell-based proxies that can be readily observed. As a hypothesis, I proposed that valve convexity (or inflation) mirrors the shell's growth rate and therefore the animal's metabolic rate.

In living bivalves that grow slowly either throughout life or in late maturity, valves are, or become, highly convex, whereas bivalves in which growth is rapid have flattened shells in which most of the shell expands in the plane of the valve's outline rather than perpendicular to it. Brachiopods also become more convex as growth slows later in life. Most living and many fossil brachiopods have strongly inflated valves, indicating slow growth and therefore low rates of metabolism. Paleozoic strophomenide brachiopods are exceptions in that their shells are strikingly flat, but there is very little space for living tissue inside the shell, again suggesting slow metabolism.

Filter-feeding sessile echinoderms such as blastoids and crinoids were also prominent members of Paleozoic marine communities. Today, crinoids are either mobile animals on reefs (order Comatulida), where they spread their filtering structures in strong currents, or sessile animals in deep water. The comatulids, moreover, have an active circulatory system. Filter-feeding echinoderms have thus either become more active, surviving in shallow-water reefs, or became restricted as less active animals owing to deep, less nutritious habitats, especially since the Jurassic.[18]

Finally, actively feeding barnacles have become abundant members of suspension-feeding communities on rocky shores as well as in deep waters and even on other animals. These bizarre sessile crustaceans are effective competitors. Their first appearance is in the Carboniferous, but they proliferated mainly during the Mesozoic and especially the Cenozoic, with acorn barnacles (Balanomorpha) rising to prominence during and after the Late Cretaceous.[19]

Gastropods are not high-powered animals, but they too have become substantially more active over time. Inspection of the shapes of the shell aperture in gastropods shows that Late Mesozoic and Cenozoic lineages developed all sorts of extensions and notches to house siphons, sense organs, and mantle fingers that construct spines. These features must all be constructed by the mantle, which extends and contracts using musculature and hydrostatic pressure, all requiring energy. Such apertural elaborations were rare except for deep slits in the outer lip of many Paleozoic and some later gastropods. These slits separate incoming water from outgoing wastes and do not necessarily involve active mantle movements. Seth Finnegan and his colleagues estimate that the average metabolic rate of gastropods has increased fivefold since the Paleozoic.[20]

At the high end of the performance spectrum, the evolution of endothermy marked a substantial increase in the metabolic rates and activity levels of animals. On land, endothermy first appeared in insects and later independently in lineages leading to mammals, pterosaurs, and birds. High blood flow into the femur of Early Permian predatory therapsid "reptiles" indicates high aerobic capacity, perhaps similar to that of large modern varanid lizards, though perhaps not yet full endothermy. Evidence from the anatomy of the middle ear of tetrapods indicates that endothermy in the lineage leading to mammals first arose about 230 million years ago during the Late Triassic. Compared with earlier animals in the clade to which mammals belong, fluid flow in the semicircular ducts of the inner ear enabled greater agility and balance, traits essential for a highly active endothermic animal. According to this analysis, the body temperatures of these early so-called mammaliamorphs was on the order of 32–34 degrees Celsius, comparable with that of modern egg-laying Australian monotremes but lower than those in most placental mammals and much lower than that of many birds.[21]

During most of the Mesozoic, true endotherms remained small and were competitively subordinate to inertial homeotherms such as herbivorous sauropod dinosaurs and low-level endothermic theropod and perhaps ornithischian dinosaurs. Warm-blooded pterosaurs, however, became predominant flying consumers of the Jurassic and Cretaceous periods, joined during the latter period by flying enantiornithine birds. All Cenozoic top vertebrate herbivores on the continents, as well as apex predators on the ground and in the air, were endotherms.

It remains a puzzle why full endotherms stayed competitively subordinate to inertial homeotherms during much of the Mesozoic. As discussed in chapter 4, even very large herbivorous dinosaurs would have had smaller appetites than later large herbivorous mammals, suggesting that plant productivity was insufficient to support a sustainable population of large Mesozoic endothermic plant-eaters. The low leaf-vein density of Mesozoic plants would be consistent with this explanation. The emergence of faster-metabolizing Late Cretaceous herbivorous hadrosaurs and predatory tyrannosaurids was contemporaneous with the rise of eudicot angiosperms with higher vein densities. The situation on small oceanic islands thus superficially resembles that of the Mesozoic, although in this case it is the small habitat area in addition to the low productivity of plants that accounts for the insular competitive dominance of ectothermic consumers.

In the sea, at least seven separate lineages of cartilaginous and bony fishes—all large-bodied predators—evolved regional endothermy, some perhaps as early as the Late Carboniferous but most during or after the Late Cretaceous. Unlike land animals, which became endothermic as small animals, marine endothermy can be maintained only in large bodies, because any heat that is produced internally would quickly be lost from small individuals with a high surface area to volume ratio. Some small diving auks and petrels (Alcidae and Procellariidae) can manage immersion in cold water by virtue of their insulating feathers.[22]

It remains unclear when fully endothermic marine animals evolved. Based on isotopic ratios and the distribution of large marine reptiles at high latitudes, claims of endothermy have been made for some Jurassic and Early Cretaceous ichthyosaurs and plesiosaurs, as well as Late Cretaceous mosasaurs, all large ocean-going predators. Even if they were only inertial homeotherms, these active animals were the top predators of their day. Large plankton-feeders remained ectothermic. Late Cretaceous birds such as *Hesperornis* and *Ichthyornis* as well as many Cenozoic birds and mammals are land-derived marine endotherms, which became top consumers—herbivores, predators, and plankton-feeders—in many ecosystems, especially at higher latitudes. Though costly to maintain in cold water, endothermy enables predators to capture slower ectothermic fish. In the tropics this advantage of endothermy is eliminated, because fish and other active ectotherms can be as fast as mammals and birds. For plankton-feeders and herbivores, however, the benefit of endothermy does not diminish toward the equator, and many endotherms with these habits are found in the tropics.[23]

It is interesting that herbivores achieved endothermy later than predators did, both on land and in the sea. There is no evidence for endothermic marine herbivores during the Cretaceous. It is only in the Early Eocene, when seacows (sirenians) entered the sea in the tropics, that marine animals became herbivorous. On land, the first potentially endothermic herbivores are Late Cretaceous hadrosaurs, but dominance by endothermic land-dwelling herbivores is a primarily Cenozoic phenomenon.

Completing the Circle

All organisms produce waste. In a healthy ecosystem, this waste is ultimately made available to the primary producers, but it takes scavengers and decomposers to convert it into products that photosynthesizers can use. If primary productivity and animal metabolic rates at the ecosystem level rose over time, rates of recycling—that is, of decomposition—should likewise have increased. Did they?

Before answering this question, I must emphasize that recycling involves more than decomposition. Many wastes in ecosystems where primary production occurs are potentially exported to systems without these producers, or even to environments where they are out of reach of any living things. To keep such losses to a minimum, wastes and the materials to which decomposers convert them must be returned to the primary producers. This requires the work of mobile animals and of rooted plants.

Decomposition on land was initially carried out by bacteria, which are still the principal decomposers under oxygen-starved conditions. Soil mixing by early animals such as millipedes and earthworms, and the evolution of deep plant roots, aerated soils by the Devonian, opening the door to fungi and animals that consume decaying plant matter. Early plants had already evolved lignin, a decay-resistant chemical complex that only some bacteria as well as basidiomycete fungi of the order Polyporales can degrade. Oribatid mites were among the earliest animals to consume dead wood, to be joined in the Mesozoic by bibionid wasps, several groups of beetles, and most importantly by termites. In modern tropical forests as well as in dry parts of Africa and Australia, termites (order Isoptera) are the most important and fastest decomposers, dating back to at least the mid-Cretaceous. All of these decomposers are ectothermic and therefore strongly temperature-dependent in their rates of consumption. The decay of plant material therefore proceeds much faster in tropical climates than in the cold. Much of this resource is rapidly and efficiently recycled under warm, well-aerated conditions, whereas it is buried out of reach of plants and animals for periods of hundreds to tens of millions of years in soils that are perpetually cold or waterlogged.[24]

The faster decay of plant leaves over time is due not only to animals with higher metabolic rates, but also to the evolution of the plants themselves—notably angiosperms—with thinner leaves. Although there are exceptions, angiosperm leaves are chemically defended by compounds that break down more easily than the more permanent, polyphenol-based compounds of early land plants as well as living ferns and gymnosperms. Their faster decay thus diminishes the loss of plant matter to sediments, where nutrients would be out of reach of the aerobic ecosystem.[25]

Besides the decomposition of plant litter by organisms, fire has been an important agency for recycling nutrients ever since vascular plants evolved. Its importance is demonstrated by the evolution of fire-resistant and fire-tolerant traits, such as pine cones that open to release their seeds only when heated by fire, and perhaps deeply hidden meristems in grasses. The intensity and frequency of vegetation fires seem to have been especially high during the Carboniferous and Cretaceous, when oxygen levels substantially exceeded modern values. With the origin of angiosperm-dominated rainforests after the Cretaceous, however, the more vigorous water cycle in those ecosystems greatly diminished the likelihood of destructive fires.[26]

Animal carcasses and dung are consumed by scavenging insects—flies, beetles, ants, and others—as well as by specialized birds such as Old World and New World vultures. In the sea this role is occupied by many Cretaceous to modern neogastropods as well as by crustaceans and incidentally by birds. Unlike plant decomposition, therefore, scavenging of animals and their excrement is done by both ectotherms and endotherms.

Economic Power in Space and Time

Two important conclusions can be drawn from the foregoing three sections of this chapter. First, the rates of all the activities of economies—production, consumption, and decomposition—have increased over time. Second, the capacity of economic power to rise has spread geographically because the most powerful consumers have become metabolically more independent of external conditions. These trends have affected ecosystems on land as well as in the sea, and have resulted in the greater integration of life in the two realms. In other words, two critical dimensions of power—speed (as indicated by rates) and size—have increased at the ecosystem level over the course of life's history.

Evidence that the power of economic systems has risen over time is extensive and compelling. The increasing intensity of bioturbation, herbivory, predation, and decomposition is coupled with an overall rise in primary productivity and metabolic rates. Other indications of increasing economic power come from the thicknesses of fossil marine shellbeds. During the Early Paleozoic, sedimentary deposits composed mainly of shells were maximally about 65 centimeters thick, one-tenth the maximum of comparable shellbeds from the last 20 million years. This increase means that secondary productivity of shell-bearing marine animals rose by a factor of about ten over the course of the Phanerozoic. Organisms have also increasingly controlled mineral cycles. Before the Cambrian, the precipitation of calcium carbonate and silica was determined primarily by physical and chemical conditions operating independently of life's activity. With the use of these minerals in the skeletons of living organisms, however, control shifted to activity in the biosphere. At first, most of this biological control occurred nearshore, but during the Mesozoic it expanded to the open ocean as well when calcareous and siliceous phytoplankters became abundant.[27]

A greater stabilizing imprint of life on planet Earth over time is also implied by an overall decrease in so-called background extinction, the disappearance of lineages during times when there are no devastating crises. This reduction has been attributed to the increasingly wide distribution of oxygen, and therefore of aerobic life, in the biosphere, a trend that is itself the consequence of more intensive organism activity in previously oxygen-poor habitats such as deep sediments and the deep ocean. Moreover, the establishment of robust feedbacks stabilizes ecosystems by making them more resistant to minor perturbations such as climatic warming, which reduces oxygen concentrations in water. Disruptions in the carbon cycle as indicated by swings in the carbon-isotopic composition of sediments and skeletons have dampened over time, and the Earth has not experienced glacial conditions at low latitudes since the Late Paleozoic. In other words, the power of ecosystems to control and regulate their own circumstances has risen over time.[28]

These historical and geographical patterns, ultimately driven by competition, expose two other historical trends in the economies of nature.

One is that, with greater mobility and distance traveled by active animals, exchanges of resources and evolutionary lineages among ecosystems have become increasingly frequent and important. Put another way, the scale of trade in goods, services, and organisms among ecosystems has become ever larger, leading to what might be characterized as ecological globalization. The second trend, correlated with all the others, is that economic efficiency—the extent to which resources are retained and used again in ecosystems and in the entire biosphere—has increased over time. A diminishing proportion of material resources is being removed from the aerobic realm of life through burial in sediments. All these emergent historical trends are powered by positive feedbacks among production, consumption, and decomposition, because these feedbacks have intensified with the evolution of increasingly powerful organisms.

Trade in Resources and Species

Consider first the increase in the scope of trade. As I discussed in chapter 5, animal locomotion transports not only organisms but raw materials as well. Sediment mixing by bioturbators makes nutrients more widely available, and vertical and horizontal movements by animals in water redistribute oxygen, particles, and dissolved nutrients. Seasonally migrating birds, bats, and insects subsidize high-latitude ecosystems with resources they bring back from the tropics, and vice versa. Birds and seals feeding at sea but nesting or giving birth on shore subsidize coastal terrestrial ecosystems with guano; and salmon, which spend much of their adult lives at sea but migrate up rivers where they spawn and die, subsidize freshwater and terrestrial habitats with marine nutrients. Land birds and mammals feeding in mangroves and salt marshes bring nutrients to marine coastal systems.[29]

On evolutionary timescales, lineages of highly competitive animals and plants have spread across ocean or land barriers and colonized ecosystems where, because of the smaller size of the recipient system, competitiveness was more constrained. A classic example is the Great American Interchange of the Late Miocene to Pleistocene, when open country mammals from North America walked across the newly formed Central American isthmus to colonize the more island-like continent of South America. Meanwhile, highly competitive South American forest birds and mammals invaded the warmer parts of North America. Similar movements from highly competitive donor biotas or habitats to less vigorous recipient ones are well documented through the geographical history of life. During the Early Cenozoic, for example, Asia supplied North America with most of the modern mammalian groups that now occupy the latter continent. During the Pliocene and Pleistocene, North Pacific marine lineages of plants and animals spread through the Bering Strait into the Arctic and North Atlantic Oceans, with very few lineages moving in the opposite direction. The mutual isolation between the eastern and western temperate Atlantic faunas through much of the Miocene ended when some European marine lineages spread westward to eastern North America during the Pliocene and Pleistocene. In our own time, species from these continents have been introduced accidentally or deliberately by humans to oceanic islands, where they have contributed to the extinction of many native insular species that were unaccustomed to the competitiveness of the newcomers.[30]

Another equally or even more consequential trade in species was between land and sea. Although dry land was likely inhabited by microbes already during the Archean eon, multicellular life originating in the sea colonized terrestrial environments during or perhaps shortly before the Cambrian period. Many lineages continued to colonize the land either from the sea or freshwater, including arthropods, annelids, gastropods, tetrapods, and plants. Some terrestrial tetrapods secondarily colonized freshwater during the Carboniferous, but no lineage is known to have returned to the sea. Marine Paleozoic ecosystems therefore acted as donors and were on the whole more powerful than recipient systems on land, as deduced from the direction of colonization. Many Mesozoic and Cenozoic animals also made the transition from sea to land, but few of their descendants became powerful agents. In sharp contrast, at least seventy lineages of tetrapods and three families of land plants became secondarily marine (not counting another thirty families of mangroves and salt-marsh plants). At least five reptilian lineages became secondarily marine in the Early Triassic, in some cases less than two million years after the end-Permian crisis. Through the post-Permian eras, many of these newly marine lineages evolved into powerful agents, including marine reptiles, whales, seals, sea snakes, birds, seacows, and seagrasses, among others. Terrestrial ecosystems after the Paleozoic, therefore, increased greatly in power and began to function as donors of species to ecosystems in the sea.[31]

Efficiency

The second, and perhaps less obvious trend that correlates with economic power in stable economies is greater efficiency over time. The power of an ecosystem ultimately emanates from a tacit, unintentional collaboration among competitors. This facilitation is an expression of the common good of the economy as a whole, in which all self-interested parties have an overriding common interest in maintaining or enhancing an economy in which resources are plentiful, predictable, and accessible. The obvious antagonism between producers and consumers must therefore be partly blunted by countervailing positive feedbacks. Both parties, moreover, depend on decomposers and scavengers to maintain the circular economy. Great power at the ecosystem scale must therefore be accompanied by strong interdependencies and to a highly efficient retention of the material resources that enter or already exist in the economy. Resources must stay in or be exchanged among living things rather than be irretrievably discarded.[32]

Some years ago, when I was teaching a graduate course in ecology and evolution, it occurred to me that a previously unrecognized contradiction exists between individual and ecosystem-wide efficiency. Powerful individual organisms such as trees, large herbivores, and apex predators produce copious material wastes and are therefore highly inefficient. Trees drop leaves, and large animals—especially endotherms—defecate and urinate copiously. High rates of photosynthesis are inevitably accompanied by rapid water loss. Endotherms shed large amounts of excess heat, and arthropods discard their external skeletons as they molt during growth. The economies in which these organisms wield power, however, are extremely efficient in that they retain almost all the material resources. Tropical rainforests, productive grasslands and savannas, coral reefs, and kelp forests are examples of materially efficient ecosystems if their complement of powerful consumers is intact. In low-powered ecosystems, where activity is constrained by scarce nutrients at low temperatures, plants and animal consumers tend to conserve or recycle wastes, but the overall system is inefficient because recycling by decomposition is slow, with the result that wastes accumulate slowly and are interred in sediments, where they cannot be retrieved by members of the system. This situation prevails in tundra ecosystems, peat bogs, and some marine and freshwater communities where the water column is highly stratified and bottom waters remain anoxic. In other words, powerful species in productive ecosystems tend to rely on replacing resources, often from outside sources or from efficient decomposition, whereas species in less productive communities rely more on retention in slow-growing, more permanent bodies.[33]

The importance of powerful organisms in making systems both more productive and more efficient is well illustrated in kelp forests and the Southern Ocean. In modern productive kelp forests of the North Pacific, the principal consumers of kelps are sea urchins of the genus *Strongylocentrotus*. As bottom-dwellers, these echinoderms primarily consume young kelps as well as other leafy and encrusting seaweeds. Left unchecked, urchins transform thriving kelp beds into much less productive barrens in which the main primary producers are slow-growing coralline red algae. Sea otters prevent this transformation by eating sea urchins, enabling new kelp recruits to grow and mature into large seaweeds. Kelps, however, evolved long before sea otters, and sea urchins were probably much less important as selective agents for kelps than were large endothermic mammals. These mammals—first Oligocene to Miocene desmostylians, then temperate sirenians (seacows) during the Pliocene and Pleistocene—were large animals that cropped kelps near the water surface, leaving intact parts of the kelp near the holdfast, where growth occurs. Among the factors that made this system of kelps and large hungry mammals so productive are fertilization by the herbivores' wastes and the thinning of the kelp canopy by grazing, which allowed light to penetrate to the seafloor where other productive understory seaweeds can thrive. The mammals thus established a strong positive feedback with their food plants. With the extinction of Steller's seacow, before the year 1750, this feedback was all but eliminated and only partially replaced by the facilitation between sea otters and kelp.[34]

The case of the Southern Ocean is even more dramatic. Calculations based on observations of feeding rates by plankton-consuming whales show that, before humans began to hunt whales on an industrial scale, 430 million tons of krill (*Euphasua superba*) were consumed by these large mammals per year. Whales were responsible for recycling more than 1.2×10^4 tons of iron, a potentially limiting nutrient for the phytoplankton on which krill feed, per year. This is 8.5 times more than the iron recycled by whales in the Southern Ocean today. Estimates of the amount of water filtered by Southern Ocean whales are 2,000 cubic kilometers before industrial whaling and only 200 cubic kilometers today. Counterintuitively, the abundance of krill has markedly decreased in today's Southern Ocean compared with populations when whales were still abundant. As a result, the Southern Ocean has become iron-limited and nutrient-rich, a low-productivity system in which recycling has greatly decreased.[35]

The human-caused elimination of many large land vertebrates, beginning in the Late Pleistocene, has seriously disrupted positive feedbacks between production and consumption and reduced productivity and stability in many ecosystems. For grasslands outside Africa, these effects are particularly well marked. With most large herbivores either extinct or greatly diminished in grasslands in Australia, North America, and South America, grasses are being cropped less effectively and are not being fertilized as much by animal dung. As a result, grasslands have become increasingly vulnerable to fires, which cause the export of nutrients from the soil and vegetation.[36]

The absence or breakdown of these positive feedbacks thus makes ecosystems inefficient. My guess is that prodigious production of coal in the Carboniferous is the consequence of low productivity of the principal plants, the absence or scarcity of herbivores, and water-logged soils in which decomposition was slow. More speculatively, if the lycophytes and pteridosperms of the coal swamps were more productive, perhaps the soils would have been less water-logged and decomposition would have occurred more quickly, resulting in more efficient recycling of dead plant litter. Herbivores would have been powerful agents of selection favoring faster growth and higher productivity in the plants. This scenario begs the question, of course, which circumstances would have prevented these relationships from becoming established. Perhaps there were hard constraints on productivity in these early plants, so that the feedbacks we have come to expect in modern or very recently extinct forests and grasslands were contingent on the evolution of angiosperms during the Cretaceous.

An increase in economic power over time implies that economic efficiency also rose. Direct evidence for increasing efficiency comes from marked decreases in the number and extent of anoxic events, especially after the Cretaceous, and correspondingly a decrease in the frequency and area of black shales and thick carbonate deposits on marine continental shelves. These deposits represent geological sinks of carbon and are therefore indications of inefficiency. The spread of oxygen, the increase of powerful agents capable of absorbing the high costs of retrieving refractory compounds that require large amounts of energy to break down into materials that organisms can use, and the production of more easily decomposable wastes by high-powered organisms all contribute to the increased efficiency of marine and terrestrial ecosystems over the course of the history of life.[37]

Diversity and Inequality

Two passive properties of economies—diversity (the number of species or occupations) and inequality (the disparity, or spread, of power or wealth among components)—have often been assigned important roles in determining how, and how well, economic systems work. Some ecologists, for example, have suggested on the basis of small-scale experiments that higher diversity (in this case of species) causes greater productivity;

and some economists blame high inequality for undesirable economic conditions and outcomes. The idea is that, as more species are added, some will exploit ways of life not present in communities with fewer species, or else one of the added species may by chance be more productive than those already present. Furthermore, there is apt to be more facilitation among species.[38]

The pioneering economist Adam Smith was one of the earliest scholars to understand that, when human societies create a predictable surplus of food produced by hunting, gathering, and especially agriculture, people who are not directly involved in food production have enough resources to fill new occupations in education, the clergy, government, the arts, manufacture, trade, and more recently entertainment. In his view, diversity of occupations can rise as an economy becomes more productive and, in my words, more powerful. Economic diversity is, in other words, a consequence of economic activity rather than a primary cause of it. Power permits, but does not necessarily lead to, diversity.[39]

The interpretation that diversity is a possible consequence rather than a cause of vigorous economic activity applies also to ecosystems in nature. Highly productive tropical rainforests and coral reefs are widely recognized as centers of high species diversity, but other productive ecosystems such as salt marshes, the pelagic zone in the Southern Ocean, and Aleutian kelp forests are not especially diverse. Moreover, there are wide variations in regional and local species numbers among the world's rainforests and among coral reefs. African rainforests, for example, are substantially poorer in species than comparable systems in South America and southeast Asia. The marine Indo-Pacific realm, comprising the western Pacific and Indian Oceans, has six to ten times more coral species both locally and in the realm as a whole than do Caribbean reefs in the western Atlantic. In the Caribbean, the highly diverse reefs of the Miocene and Pliocene are smaller and patchier and grew more slowly than the less diverse Pleistocene and Holocene reefs.[40]

Instead of being a catalyst of production, diversity in ecosystems seems generally to increase in the absence of substantial disruption. Among major marine geographical regions, this relationship is evident in both the temperate and tropical zones. In the modern marine temperate biota, the North Pacific is by far the richest, and has been least affected by extinctions during the Pliocene and Pleistocene. For the tropics, the Indo-Pacific realm is the richest in species, and although there have been Pliocene and Pleistocene extinctions there, these have been fewer than in other parts of the tropics. The overall increase in biological diversity over the Phanerozoic may thus reflect the increasing long-term stability of the biosphere rather than it being a direct consequence or cause of increasing power.[41]

Every ecosystem, and likely every species, is characterized by inequality among its members. Inequality in wealth (biomass) or income (resources made available per unit time) can be described as a power law, where a few members hold a disproportionately large amount of power and most members have little. The power law applies at all scales, from individuals in a society to species in an ecosystem to nation-states on the political map of the world. One expression of inequality is in the hierarchical organization of a system: the more hierarchical (or stratified) a society or ecosystem is, the larger is the difference between the most and the least powerful members, or the proportion of resources controlled or earned by elites. Mobile hunter-gatherer societies, for example, are substantially less unequal in the distribution of wealth and income than are sedentary, more territorial societies in which more of the wealth can be passed from one generation to the next. Put another way, if power can be inherited, it has a tendency to accumulate and to become more concentrated. This therefore exacerbates inequality. If, moreover, competition among powerful members is an agency favoring even more power, there will be a long-term trend through history of increasing inequality. At this large scale, therefore, inequality is an expression and an outcome of increasing power at the top.[42]

A pressing concern in modern human societies is the excessive concentration of wealth and power vested in a very small number of individuals and corporations. I shall have more to say about this problem and possible solutions to it in chapter 8. In the ecosystems of nature, the only generally applicable mechanism for reducing inequality and diversity seems to be extinction, which disproportionately targets the most powerful players.

The Economics of Extinction

If economic stability and power depend on a reliable supply of raw materials either from outside the system or through recycling, then disruptions that interfere with production and with the collaborative feedbacks among sectors will compromise the system's power and efficiency. The history of life chronicles catastrophic crises, usually separated in time by millions of years, which severely disrupted economic processes. The most obvious manifestation of these disruptions was mass extinction, during which many species and lineages disappeared more or less simultaneously. Although mass extinctions have usually been studied from a taxonomic perspective, it is their ecological repercussions with which I am concerned here.

Despite their differences, the great mass extinctions of the past have in common the substantial loss of primary producers, plants on land and phytoplankton in the ocean. Regardless of whether the ultimate causes of these extinctions are collisions between Earth and a celestial body or extraordinarily intense volcanic activity, the loss of primary production is likely to be the consequence of the release of vast amounts of sulfur, smoke, and dust into the atmosphere, shielding Earth from sunlight and exposing life to a nearly global period of toxic emissions. Large active animals, which are more dependent on a predictable and abundant food supply than other consumers, would be among the most vulnerable members of ecosystems under these circumstances. A supply-side crisis therefore also has a strong, correlated demand-side component. Before the end-Permian crisis, for example, several large marine and terrestrial animals weighed as much as one to two tons. All these giants disappeared during the crisis. At the beginning of the succeeding Triassic period, the largest land animals weighed not more than thirty kilograms. Even more dramatic reductions are observed after the end-Cretaceous extinction event.

Paleontologists have long known that the mass extinctions of the end-Permian and end-Triassic, as well as some less severe events during the Paleozoic and Mesozoic, are associated with widespread anoxic conditions even in shallow marine waters penetrated by sunlight. Anoxia is inimical to aerobic life and has often been blamed for causing the extinctions of animals. Contributing to such deaths is intense ocean acidification, which has also been documented for the most dramatic mass extinctions. Given that nutrient supply and demand are intimately linked, however, it is entirely possible that anoxia and acidification are consequences of economic collapse rather than primary causes of extinction. They certainly could have contributed to species losses, but they are best interpreted as manifestations of economic collapse.[43]

Indications are that initial recovery from extinctions was geologically rapid, especially for tiny phytoplankters with short generation times and life spans. After the 50% drop in marine primary production during the end-Cretaceous crisis, phytoplankters recovered to pre-crisis levels within at most 30,000 years. Some mud-dwelling animals at the site of asteroid impact in the Chicxulub crater on the Yucatan peninsula of Mexico had recolonized the seafloor at about the same time, as indicated by trace fossils. Elsewhere, occupation of the seafloor was much slower, and it remains unclear when stable links between primary producers and consumers were reestablished. On land after the Cretaceous, it took 300,000 years for mammals as large as 20 kilograms to evolve, and 700,000 years before a mammal weighing 47 kilograms appeared, as demonstrated by a fossil sequence in Colorado. Earlier, after the end-Permian catastrophe, recovery seems to have been slower, possibly because physical conditions remained unstable for several million years. High productivity was evident in the ocean in Oman immediately after the crisis, but marine communities with powerful consumers such as early ichthyosaurs and other marine reptiles appeared two million years after the Permian. The first marine apex predator appeared six million years after the crisis. In short, stabilizing and positive feedbacks took millions of years to be reestablished.[44]

Despite the disastrous effects of mass extinctions, the aftermath of the crises set the stage for the evolution of traits that would later prove to be critical for the establishment of new, more powerful lineages and a new world order. In particular, initially there must have been a premium on rapid growth, enabling surviving organisms to capitalize on poorly regulated but potentially plentiful resources. The high metabolic rates required for rapid growth would be prerequisite for the evolution of larger organisms with longer life spans and more competitive staying power. Even if those more escalated lineages took millions of years to evolve, I suspect that their ability eventually to surpass the performance levels of pre-crisis counterparts was made possible by the rapid growth of surviving organisms immediately after the crisis.[45]

In the post-Cretaceous history of terrestrial ecosystems, for example, early recovery featured small, fast-growing mammals. Large size came later—often much later—and enlargement of the brain relative to the rest of the body also began well after the end-Cretaceous extinction. From the Middle Eocene onward, first predatory mammals and then herbivorous ones began to expand the parts of the brain involved in vision, posture, and hearing at the expense of the older parts related to olfaction. More overt interference competition is thought to be responsible for this relative brain increase, replacing the emphasis on only rapid growth and early reproduction. The increase in brain size, also observed in primates after the Eocene, set the stage for the origin of hominins in the Late Cenozoic.[46]

Economic Vulnerability from Within

Power confers many advantages to economic systems, including a degree of independence from external agencies, a greater capacity to retain and reuse resources, and the ability to recover from great crises. Great power at the ecosystem level could, however, also usher in an increased threat of internal disruption. If over time the constraints on individual and collective power are relaxed, then the possibility that some sector will acquire truly exceptional power also becomes more likely. In other words, it could be that internal sources of destruction intensify even as the influence of external agencies wanes. Like outside forces, internally generated disruptions could modify or even compromise the collaborative arrangements on which the acquisition of great power depends.

In the history of life until the Anthropocene, internal disruption directly caused by organisms seems to have been at most a regional and temporary phenomenon. An argument could be made, for example, that the arrival of North American mammals in South America brought about substantial disruptions in recipient ecosystems, perhaps because the perpetrators are endothermic mammals. Comparable invasions of mostly ectothermic organisms from the North Pacific to the North Atlantic by way of the Arctic certainly modified recipient Atlantic marine ecosystems, but did not bring about the extensive extinctions of native species that occurred in South America as a result of the Great American Interchange. The history of internal disruption has not been thoroughly investigated in the geological record, but all the past mass extinctions and probably most lesser events of species loss are attributable to external phenomena such as warming, cooling, sea-level change, volcanic eruptions, and meteorite impacts. Although life may have influenced some of these changes through feedbacks between climate and activity, the principal causes remain geological and celestial.

With the arrival of our own species, however, this situation has dramatically changed. In the next chapter I shall argue that, although our evolution represents a continuation of past trends in individual and collective power, the human economic monopoly has introduced a novel threat to the biosphere. The circumstances under which this was able to happen and what might be done about self-limiting our power are topics about which the long-term history of life has potential insights to contribute.

8

The Human Singularity

The evolution of technologically advanced *Homo sapiens* is one of the few truly singular events in the history of life. No species before us has wielded anything close to as much power as our species, either individually or collectively, and no single species has so profoundly altered planet Earth geologically and biologically. We are the first species to create an entire economy, with all of the functions that in the rest of the biosphere would be performed by multiple species. This novel economy, built on top of the remaining biosphere, has in effect become a one-species monopoly, a system with unprecedented power and correspondingly great internal vulnerabilities.

This development raises many important questions about history and the future. Are the circumstances that enabled our rise to power as unique as the economy we have built? Does our economic ascendancy represent a continuation of previous trends, or is it a fundamental and thus unpredictable departure, ushering in processes and outcomes that have no precedent? Is there a global limit to human power and, if so, how would such limits be imposed? Can humans collectively construct an economic system that does not grow but that is healthy and beneficial to all its members? What would such an economy look like?

The argument I present in this chapter is that, though unique in many respects, humans and the economies we have built follow the same principles and trajectories toward greater power that have characterized the biosphere since life began on Earth. The time course of our history is vastly shorter, and our spatial reach much greater, than that of previous trends, but the fact that they follow the same course reflects the underlying unity of agency as expressed in competition and in the establishment of essential cooperative relationships. The problems that excessive power present are unusual but not unique in the history of the biosphere. How the biosphere dealt with such earlier challenges could provide clues about our own attempts at a sustainable and regulated economy. I suggest that, whereas in nature economic health and durability are established and maintained diffusely, they will likely require deliberate policies enshrined in laws and norms enforced by governments for the common good of the whole biosphere in the human-dominated future.

What Makes Humans Unique?

It is beyond doubt that the human species *Homo sapiens* is a primate mammal, part of the hominid clade of anthropoid great apes that diversified as facultatively bipedal animals beginning during the Late Miocene in Africa, about seven million years ago. Our genetic makeup is indeed remarkably similar to that of our closest ape relatives, the chimpanzee (*Pan troglodytes*) and the bonobo (*P. paniscus*), strongly implying that our distinctive characteristics cannot be attributed to our genes alone. The unusual physical features and abilities that set the genus *Homo* apart from other great apes may be prerequisites or corollaries of our rise to preeminence, but by themselves they cannot account for the evolutionary trajectory of our species. These physical traits include adaptations for endurance running, such as hindlimb and pelvic modifications; shoulder and forearm adaptations for accurate throwing of objects; and the migration of the base of the tongue to the throat, enabling adult humans to shape and diversify sounds in the mouth as part of spoken language. Even our most celebrated structure, the large brain with its expanded frontal lobes, evolved 1.7–1.5 million years ago, long after the origin of *Homo* but much earlier than the modern globular shape of the brain had become established some 50,000 years ago, and well before humans individually and collectively achieved biospheric hegemony. Humans also evolved higher metabolic rates and more body fat than other great apes, likely in concert with the enlarging, energy-demanding brain. All these anatomical changes offer great potential, but by themselves they represent unexceptional adaptive evolution. The realization of that potential resides in the emergence of a novel social order. In fact, elevated internal metabolism likely correlates with early social and cultural innovations such as cooking and sharing food, an increased reliance on high-energy meat in the diet, and the use of long-distance weapons for killing large prey.[1]

The link between physical and cultural attributes deserves emphasis. Our anatomy reflects adaptive responses to our environment, especially its biological components, but it is also inextricably bound up with agency, that is, with what we do individually and collectively. An intriguing, somewhat speculative example will reinforce this point. With the beginning of agriculture and invention of implements to cut food into smaller pieces outside the body, foods tended to become softer, requiring less force to be applied by the teeth. The juvenile configuration of having the upper front teeth slide in front of the lower teeth persisted into adulthood with this new regime, replacing the adult Neolithic condition in which the upper and lower teeth met edge to edge. This culturally induced anatomical change in bite made available novel speech sounds formed by the lower lip meeting the upper teeth. As a result, the repertoire of sounds (phonemes) expanded in many languages. The important point is that agency, or behavior—including learned behavior—is instrumental in the evolution of anatomical traits and configurations. More generally, the mechanical forces that agency unleashes affect the ways in which muscles, bones, teeth, and even many internal organs develop, as anyone who regularly exercises can confirm. Moreover, these force-induced changes often come under stronger genetic control.[2]

The key to our power as individuals and as a species is the formation of coherent groups that act as individuals and that have culturally transmitted characteristics. Social organization has a long history among animals in general and among primates in particular, including nonhuman great apes, but only in *Homo* is there a tendency toward what has been called ultrasociality. This rare form of social organization entails extensive division of labor among individuals within a group as well as among groups. These groups and the larger institutions they form are coherent entities by virtue of shared beliefs, practices, norms, and goals, which provide the social glue for genetically unrelated, self-interested individuals to work together for the benefit—or common good—of the group. Importantly, and uniquely among animals, humans communicate via spoken (and later written) language, an open-ended, combinatorial system analogous to the genetic code of organisms. Language enables individuals and groups to create, gather, accumulate, and transmit knowledge that informs deliberate actions and decisions.

The emergence of ultrasociality requires that an individual's likelihood of survival and leaving offspring increase even as individuals cede some of their self-interest and autonomy to the common interest of the group. Coherent social groups should thus be competitively superior to individuals acting alone because they can accomplish work that individuals by themselves cannot and because they offer benefits of safety, mutual aid, and shared resources. Some of these advantages accrue to any social group, but human organizations are distinctive in developing culturally heritable traits. They are held together by rituals and sanctions embodied in laws and religions and reinforced by allegiance to a charismatic leader or an invented supernatural authority. By contrast, most animal societies such as flocks of birds or schools of fish are temporary entities without heritable characteristics or a shared ideology.[3]

The concept of the common good is, of course, not unique to modern human society. From its very beginning, the history of life has been characterized by cooperative and mutualistic arrangements in which initially autonomous agents come together to form a large whole, which it is in the interest of the members to perpetuate. Organelles like plastids and mitochondria in cells are kept in reproductive check by their greater common interest to maintain and propagate the cell in which they live. Symbionts in larger organisms likewise give up their selfish reproductive imperative in favor of the survival of the host; and in ecosystems, cooperative arrangements among would-be competitors enable a system of self-interested entities to provide the services and stability that all members need.[4]

Unity within human groups, and likely within most social groups of organisms, is in many cases forged by the presence of a real or perceived common enemy. Such cohesion, which is especially well exemplified by military organizations, makes possible and is promoted by intense inter-group competition. Samuel Bowles in fact has suggested that warfare and within-group cooperation have culturally evolved together, and that this feedback between cooperation and competition could have helped propel our species to ultrasocial organization. Some group conflict is known also in ants and chimpanzees, but it has obviously reached far more destructive levels with the development of increasingly powerful weapons.[5]

Humans are exceptional among animals not merely in having developed an ultrasocial organization in which groups wield great power and group membership is of paramount importance, but also by preserving and indeed celebrating the accomplishments of individuals. Writers, inventors, artists, as well as many leaders, scientists, and entrepreneurs, work alone and are recognized (or criticized) for their work. They operate in society, of course, and whether their work is accepted or rejected depends largely on the social context, the marketplace of ideas. Even in the absence of notable accomplishment, every individual has a name and an identity, a distinctive personality with preferences, likes and dislikes, a role to play in society, and perhaps above all a reputation. Societies differ in the degree to which individual initiative and independence are valued, but in no human society is individualism wholly subordinate to group identity.

Individuals in human societies are not interchangeable, but instead are considered distinct from each other in ways that matter socially. It seems very likely that mate selection in animals, where individuals choose or compete for potential mates as individuals, provided the ancestral condition from which our exceptional individualism evolved. Nevertheless, it is more than a little ironic that our species should have simultaneous tendencies toward ultrasocial and unique individual behaviors.[6]

Whence Human Uniqueness?

As I implied in the preceding section, the evolution of human ultrasociality incorporates an element of self-perpetuation thanks to feedbacks between anatomy and culture, energy and technology, and competition and cooperation to establish an economy. Given that the emergence of modern humans is a unique event in the history of life, it is worth asking how and where such a self-perpetuating series of changes came about. Under which circumstances did the initial stages arise, and could they have existed elsewhere and at other times, or are these initial conditions unique?

That the genus *Homo,* and apparently our own species *H. sapiens,* evolved in Africa provides important information about the circumstances in which both our physical and social traits originated and evolved. Africa was and remains a highly competitive place with fierce predators, large and aggressive herbivores, fast, venomous snakes, and a large complement of parasites and their vectors. In addition, social anthropoid primates had existed there ever since the Late Oligocene, about 25 million years before present. Many of these primates, including our immediate hominid ancestors, were not physically fierce, lacking such defenses as sharp claws, horns, tusks, and a very large body. Instead, they took advantage of collective action in procuring and defending resources, and augmented these group benefits with weapons fashioned from sources outside the body that could be deployed at a distance. Competitive rivalries among small bands led to collective combat, cementing the cultural bonds within groups and thus making them more cohesive.[7]

Conditions favorable to the evolution of ultrasocial primates might have existed on other landmasses as well, but there are reasons to believe that few if any continents could have matched Africa as the most promising birthplace for a species with the potential to exceed the power of any other living entity on Earth. In terms of area, Africa is rivaled only by Eurasia and North America among modern continents, especially because these two huge landmasses were connected across what is known as Beringia during much of the Cenozoic. Africa, too, was and is connected with Eurasia, enabling mammals and other species to move between these continents at various times over the last 45 million years. Many of Africa's mammals—ultimately including hominids at various times—found their way to Asia, and Asian herbivores such as bovids with well-developed head gear flooded into Africa during and after the Miocene. Eurasia therefore might have been just as amenable as Africa to the evolution of socially and technologically advanced hominids, and it is indeed possible that some attributes of modern humans developed in Asia. North America also supported a large complement of powerful mammals, but although it was highly receptive to the arrival of modern humans from Asia during the Late Pleistocene, suitable social ancestors that could have given rise to even more social humanlike species were absent on that continent. The other continents were too small to have supported the highly competitive mammalian faunas of the larger landmasses, although all of them—South America, Australia, and Madagascar—proved receptive to colonization by high-performing mammals from the large continents.[8]

Once the initial evolutionary steps toward ultrasociality had been taken in Africa, subsequent cultural evolution and innovation occurred independently and in parallel in many parts of the world where humans had colonized, and therefore became much less contingent on the unusual coincidence of enabling factors and selective agencies on that continent. Crucial cultural innovations such as animal and plant domestication, urbanization, nation-states and empires, religions, corporations, and industrialization transformed many independent societies. It is true that, as Jared Diamond compellingly argued, southwestern Asia and the Eurasian steppe were unusually well poised for the development of agriculture, because animals with relatively benign dispositions and high-yielding cereal plants were diverse and available for domestication. The animals not only served as sources of milk, meat, wool, and hides, but they could also be employed to work in planting and transport, and even in warfare. It was also in Eurasia where the wheel was invented. Nonetheless, agriculture arose elsewhere too, and urbanization, hierarchical societies, religions that united disparate people into coherent and powerful groups, and elaborate expressions of culture in the forms of art, architecture, and ritual arose, whereas populations grew large enough and resources were sufficiently plentiful. These cultural developments, in other words, took on a life of their own. It is therefore particularly puzzling that modern science, which powered humanity to monopoly, appeared so late and in only one society, that of Western Europe.[9]

The origins of modern empirical science are entirely cultural. Intellectual leaders have existed always and everywhere in human societies, but only in the seventeenth and eighteenth centuries in Western Europe (especially England and Scotland) did they grasp, and help to make socially acceptable, the idea that knowledge—facts and insights drawn from those facts—can be most effectively gained by observation and experiment without intrusions from theological doctrines and philosophical thought. The development of scientific inquiry coincides with many other cultural shifts, including the rise of the nation-state instead of religious authority as the primary unifying social institution, and to my mind also the practical requirements associated with worldwide exploration undertaken by Europeans. As a consequence of these explorations, Europeans were confronted with a barrage of novelties—foods, different human societies and customs, climates, and more—that necessitated, or at least encouraged, a rethinking of old established thought inherited from ancient authority. Michael Strevens, who has studied the scientific revolution from a philosophical perspective, points to the compartmentalization between the demands for empirical inquiry and the more philosophical and religious humanist dimension of thought. According to Strevens, such compartmentalization would have been unthinkable in any society before the seventeenth century because the humanist perspective was so thoroughly integrated—and part of any explanation—in the minds of natural philosophers. Whatever its origin, the intellectual evolution that led to modern science was crucial to the development of the human monopoly over the biosphere.

Building the Human Monopoly

Throughout human history, individuals as well as coherent groups have competed fiercely for both material goods and status, but they were also engaged in far-reaching cooperation in production, trade, manufacture, crime, the arts, education, warfare, and government. Collectively our species has constructed an economy and built cities and transport systems. This economy, complete with extensive division of labor, has grown in power, raised living standards, extended life expectancy, and harnessed scientific knowledge to overcome long-standing limits on population size, food production, and the time and energy devoted to labor.

These extraordinary accomplishments were made possible by many economic innovations, from money as a means of exchange to banking and credit, for-profit corporations funded by investors, the concept of insurance against disasters, and centralized government. All these inventions have in common the ultrasocial emergence of trust, the expectation that people will honor commitments of cooperation and help by institutions despite diverging self-interests of the parties involved. Trust is the glue that enables mutual strangers to cooperate in trade and more generally in the economic marketplace. Money could not function if people did not agree to trust its value when they exchange one good or service for coins or bills that by themselves have no value but that can be used to buy something else. Credit and debt, facilitated by banks and other businesses, permit investment and trust in future activities, and together with insurance spread the risks of failure. Government cannot operate effectively without a substantial part of the citizenry trusting it to act in the society's interest. Science, the organized system of acquiring and disseminating verifiable knowledge, depends on widespread public trust in its methods and findings as well as on the expectation that it, too, will inform policies beneficial to society.[10]

Society dissolves into chaos and becomes vulnerable to authoritarianism and demagoguery when, as a result of self-serving manipulation by cults and charismatic leaders, people lose trust in institutions that rely on verified knowledge and reason. Given their ultrasocial nature and their urge to identify with a group, people are eager to extend trust, especially to a parochial circle of people they know. Regardless of where trust is placed, it entails a leap of faith, a hopeful acceptance of a social contract. It may matter little to people if that contract is based on a fictional belief system enforced by advertisement, propaganda, or a religious authority, or if it is built on an empirical foundation. What seems to matter above all else is the ultrasocially induced satisfaction of trusting, belonging to, and identifying with one's tribe.

The evolution of institutions that elicit trust requires communication. As the spatial scale of human activity and communication expanded, so did the size of institutions that help spread risk. With increasing trade among trusting strangers, human activity became more and more global. Previously isolated societies came into contact, exchanged crops and belief systems, and learned from each other about worlds unfamiliar to them. None of this could have happened without trust, but neither could it have transpired without the invention and acceptance of novel energy sources, labor-saving devices, long-distance transport and communication, and above all in novel institutions such as limited-liability companies, banks, and insurance. Trust, invention, and social acceptance of a new and rapidly changing world order were necessary ingredients—and consequences—of the meteoric rise in human economic power and the ascent of the human monopoly.

Nothing sets modern humans apart from other life-forms more than the extraordinary power wielded by our species both as individuals and collectively. Figures for energy consumption shown in table 8.1 demonstrate a dramatic and strongly accelerating rise over the last 11,700 years, with the last 70 years accounting for more energy than the preceding 11,000 years. This pattern reflects an increase in the number of energy sources—from fire to domesticated animals, water, wind, coal, petroleum, and nuclear fuels—as well as a remarkable rise in the human population by a factor of almost 1,950. Since the year 1750, just before the Industrial Revolution, the use of fossil fuels has risen by a factor of more than 800. Even over the short span of the Anthropocene, there has been a fifty-fold increase in fossil-fuel use. Jan Zalasiewicz and his colleagues estimate that the mass of physical structures, tools, machines, and wastes accumulated by humans amounts to a staggering 30 trillion tons, outstripping human biomass by five orders of magnitude.[11]

The acceleration in human power and population size is closely paralleled by the pace of major innovations and inventions. From long-range weapons and symbolic thought to agriculture, urbanization, writing, printing, and the computer, these cultural innovations have speeded up and increased the reach of every function in the human economy. As shown in table 8.2, all these accelerants involve either the tapping of previously unused energy sources or social and technological inventions, or both, which are also external to the human body. On the darker

Table 8.1. Energy use per year and population size of humans during the Holocene and Anthropocene

Date (years before present)	Per capita energy use (109 joules)	Collective energy use (1021 joules)	Population size
11700–8200	5.8–6.5	0.12	4–8 million
8200–4300	6.5–7.8	0.4	8–27 million
4300–350	7.8–40	1.42	27–2500 million
350–170	13.5–22	2.9	600–1247 million
170–70	22–40	4.9	1250–2500 million
70–0	40–75	22	2500–7800 million

Table 8.2. Dates of origin of human social and technological inventions

Invention	Date (years before present)
Agriculture	12,000–10,000
Wheels	6000–5000
Urbanized society	5700
Empires	4400
Use of iron	3100
Coins as money	2600
Wells	2500
Windmills	800
Firearms	670
Printing	570
Precision tools	240
Artificial fertilizer	235
Airplanes	120
Plastics	100
Nuclear power	75
Computers in wide use	40

Source: Data modified from Vermeij and Leigh (2011).

side, the power of weapons has witnessed a comparable or even greater acceleration in power and innovation to the point where humans are perilously close to destroying the entire aerobic biosphere.[12]

All these developments have transpired over a remarkably short time, with the most dramatic increases taking place during the last 250 years.

Nevertheless, they exhibit the same pattern of increasing power and accelerating change as does the ecological and adaptive history of life as a whole, except that the pace of prehuman history is slower by some five orders of magnitude and is controlled to a much greater extent by genes. The fact that the directions of change are the same in the human and nonhuman realms implies strongly that the same agencies of competition, cooperation, and adaptation are at work, regardless of how adaptive characteristics are transmitted and accumulated.[13]

Full domination of the biosphere by humans had certainly become undeniable by the beginning of the Anthropocene, but there is growing evidence that human activity has substantially affected our planet for much longer. Through a combination of hunting, a greater frequency of human-set fires, and some deforestation, nearly all large mammals in North and South America became extinct, as did many in Eurasia and Australia. With the beginning of agriculture and pastoralism, humans wrought significant and accelerating landscape changes. Even during these early stages, the human stamp on the Earth already exceeded that of any previous single species. The trajectory toward human hegemony was therefore well under way even before urbanization, to say nothing of subsequent developments leading up to the Anthropocene.[14]

As in nature, the rise in power of individuals and of human society as a whole has reduced the vulnerability of the economy to external shocks. Early in our history, disasters like famines, pandemics, earthquakes, and volcanic eruptions were frequent causes of high mortality. Life expectancy was low, and adding more people to the population worsened the plight for all in what has come to be known as the Malthusian trap. Reduction in the severity of all these hazards came with advances in public health and sanitation, agricultural practices, food preservation, mechanization of labor, the rise of the welfare state and, beginning around the year 1850, the rise of science and medicine. These developments greatly benefited human society, but they also eliminated controls and constraints on human numbers and activity, with the result that nonhuman agencies no longer stood in the way of the ascent of humans as the dominant power on Earth.[15]

The Perils of Monopolistic Power

The fundamental condition of a monopoly is that there is no effective countervailing power to limit it. Economists have generally criticized monopolies for distorting the so-called free market, which through interactions between buyers and sellers is supposed to regulate prices of goods and services without overt interference from government. In reality, if one side—typically the seller—is much larger and more powerful than the buyer, the price is distorted in the seller's favor, and the market is dysfunctional. Power is rarely equal in the interacting parties, with the result that markets are more often distorted than fair. Only when rules and regulations limit power by establishing cooperative feedbacks can a market be expected to work to the advantage of both sides. This is why monopolies are rare in nature: Consumers depend on producers and vice versa, and although consumers typically have more power than primary producers, their power is limited by the diffuse regulations of the ecosystem in which they work. The free market, in other words, is an illusion, a construct existing in the mythical world of perfection. A sustainable economy requires the regulated distribution of power. Limits are imposed by what might be termed fairness feedbacks, which arise as the result either of innumerable interactions among adapted organisms or of explicit agreed policies and laws enforced by government.[16]

I contend that the perils of monopoly extend far beyond the relatively tame problem of unfair pricing. An excessive concentration of economic and political power, together with the expectation for continued growth of the monopoly, comes with four great vulnerabilities: the dissolution of trust, already alluded to above; the erosion of mechanisms to correct mistakes made by the monopoly; the tendency to equate or confuse the good of the whole with the good of the powerful; and, above all, unregulated environmental destruction that has the potential to destroy the biosphere on which we and the rest of life depend. If these dangers are to be diminished or reversed, they must first be acknowledged and diagnosed, a step that until now has been taken only hesitantly and tentatively and then only by a small number of people. We have progressed little, metaphorically speaking, from the age of toddlerdom.

Unopposed authority or power can in principle be effective and rapid in bringing about beneficial change, but that possibility hinges on whether the change is in fact the right one under the circumstances. Even good decisions have potentially harmful unintended consequences, almost ensuring that uncontested policies stand a good chance of doing great harm. Without alternative solutions (or hypotheses) being considered and tried (or tested), mistakes cannot easily be corrected. China's Great Leap Forward of the 1950s and the Cultural Revolution of the 1960s and early 1970s are prime examples of these hazards; so were Napoleon's and Hitler's decisions to wage war against formidable opponents on the latter's own soil. All were motivated by mistaken dreams of national glory. Such hubris, when not tempered by well-informed qualification and opposition, rarely ends well. In effect, monopolistic power offers a poor recipe for crafting, evaluating, and implementing solutions to real or perceived problems. It creates a world without effective adaptation.

Contributing to the possibility that groups—especially monopolies—can make uncorrectable mistakes is the ease with which false, misleading, and often malicious information can spread. This problem exists in nature as well, as the spread of cancerous cells makes plain, but it is a much greater threat when information is spread culturally. Propaganda, unjustified hero worship, unsubstantiated rumors, and prejudicial demonization of outsiders are as old as human history itself, but instant communication via social media, largely operating without editors and other meaningful gatekeepers, has allowed unverified and simply wrong statements to spread unchecked, with the pernicious effect of eroding trust. Uncritical acceptance blurs the distinction between verified knowledge and nonsense.

The problem isn't only that harmful or false ideas propagate easily and find fertile ground, but that society's resistance to them is increasingly ineffective. In our ultrasocial species, reception of ideas depends heavily on whether we trust their source and admire the people who accept them. Self-serving propaganda might fail if trusted friends and authorities are skeptical of it. In Weimar Germany, for example, Nazi ideology was vigorously opposed by the Catholic clergy, with the result that fewer Catholics voted for the Nazi party than did Protestants, whose leaders were more sympathetic to Hitler's Fascist ideology. If the social order works well for people, as for example when honest government provides consistently effective services, demagogues and liars might gain less traction than if those in power are corrupt and sow discontent and mistrust. Our emotional state affects how we perceive and react to the world, and influences what we accept as truth. Modern human society, in which interactions among people are by far the most important determinants of cultural selection, may be especially vulnerable to malevolent actors, because the environment to which we respond has gradually shifted away from the challenge of nature to the demands of operating in society.[17]

Perhaps the human receptivity to misinformation is not surprising. As one mechanism for establishing cohesion in ultrasocial groups, humans have created stories, myths, and doctrines that often lithify into unquestioned holy texts whose content contradicts knowledge gained through methodological inquiry. Like fairy tales and other fanciful fictions, those myths are part of every child's upbringing. Moreover, because they typically incorporate moral lessons, they remain influential in the lives of many adults. Humans are, in other words, primed from an early age to take counterfactual stories and concepts seriously. Without explicit training in, and respect for, critical and rational thinking, it is challenging and often socially undesirable for people to jettison the fictional dimension of their identity.[18]

Equating the good of the whole with the good of the powerful has been a widely recognized peril of excessive power. Adam Smith famously argued that the economic activities of individuals would automatically—by means of the invisible hand—benefit the larger society even when individuals do not have those benefits in mind. As David Sloan Wilson and John Gowdy have pointed out, however, the invisible hand was never much more than an expression of hope. It can operate at an unconscious level only if the group also benefits the individual, that is, when there is a positive feedback between the individual and the group.

Smith certainly did not consider how the invisible hand might fare in the grips of a monopoly.[19]

The concordance between self-interest and group interest is violated in societies in which power is excessively concentrated. The most commonly recognized violation is the tragedy of the commons, in which individual self-interest eclipses the good of a poorly organized—and therefore relatively powerless—collective whose properties are taken for granted by individuals. Typically, this tragedy takes the form of individuals appropriating resources that are held in common by the group without regard to the depletion of those resources. In his original formulation of this idea, Garrett Hardin did not specify which kinds of groups or interests would be most at risk. This is why I have added to Hardin's characterization the caveat that groups are not well organized, so that they do not act as individuals. People acting alone may neither recognize nor care about the harm they inflict on the commons. This seems to be the case, for example, when the collective suffers pollution, deforestation, overfishing, or climate change. Moreover, if the harm appears to be far away or far in the future, short-term advantages will loom larger than long-term or distant harm. Such incongruences crop up when investors force companies to maximize profits in the short run over more sustainable activity in the long run.[20]

There is a second related concept, which I have called the tragedy of the individual. In this case the perceived good of the group supersedes the self-interests of individuals. This happens, for example, when groups wage war to the detriment of soldiers and civilians. The doctrine of eminent domain, where a government or corporation seizes the property of individual owners for a public project, is another example. Unless individuals organize against the group—hard to do when there is a war on—the group interests prevail.[21]

The tragedy of the individual exposes a troubling myth about groups. It has become fashionable to speak of the wisdom of crowds, the idea that decisions made collectively are often better than those made by individuals. It is indeed surely the case that a consensus reached by a group can be broadly beneficial, but the contagious spread of ideas and emotions—good and bad—ensures that there is no guarantee that a group's action is always wise. Mobs, rebellions, and mass hysteria are well-known examples of group activity that can go terribly wrong. The stupidity of crowds is as real as is collective wisdom, and is both a symptom and a cause of group domination over the interests of less powerful individuals.

In each of these highly asymmetrical interactions in a monopoly, the interests of the more powerful entity prevail against those of the weaker party. Most importantly, adaptation by the weaker side is difficult or impossible owing to the large power differential. The absence of a significant positive feedback between the powerful and the powerless may stem from the monopoly itself. In the absence of competitors, a monopolistic entity has no incentive to establish a benefit to the weak. The rules that benefit the monopoly are created by it, and make it possible for the dominant entity to acquire even more power, expanding and deepening the economic and political chasm.

The phrase "good of the group" raises some additional interesting problems and troubling questions that have remained surprisingly unexamined. What is good for a group at one level in a hierarchy of groups may not be good for the larger umbrella entity. Some countries, for example, will benefit in the short term by burning coal, but this benefit does not extend to the whole world. In other words, the tragedy of the commons can be scaled up, where a small group takes the place of an individual in some of its interests to the larger body. A related problem is that, with increasing group size, the diversity of component groups with diverging interests will also rise, potentially compromising the larger group's cohesion. This trend could be counteracted by the establishment of a hierarchical structure, which unifies a multifaceted collective but also makes it behave more like a monopoly. Moreover, as the scale of a group expands, its common interests might become more remote to its members, especially those in the lower reaches of the hierarchy. Anyone working in a large bureaucratic institution such as a university, government, or giant corporation can appreciate these effects.

These problems are not trivial. Throughout human history, there has been a pervasive tendency for group size to increase, from small tribal groups to nation-states and empires, from small schools to large universities, from villages to megacities, from local cults to global religions, and from small enterprises to corporate behemoths. Each of these increases represents not merely an increase in power, but importantly a concentration of power in the hierarchy of power. Conflicts among the parties comprising the large supergroup can render the whole more vulnerable to internal disruptions, threatening the structure's power and persistence. As I already noted in chapter 1, these internally generated instabilities could account for the relatively short duration of some large entities such as empires and monopolistic corporations. Other large entities, such as universities, have more clearly defined single functions and may not be as vulnerable to internal disruptions. The "good of the group" is better defined and more obvious to its members when institutions and interests are well delineated than in cases where too many divergent interests are accommodated under one wobbly umbrella.

Like many others, I see environmental degradation as the greatest threat confronting our species and the biosphere. Economists and political leaders are nearly unanimous in their expectation and hope that the economy can continue to grow, even as they acknowledge the harm of overfishing, pollution, deforestation, habitat fragmentation, ocean acidification, and human-caused climate change. Much less appreciated, but at least as important, is the increased instability of remaining "natural" ecosystems now that we have eliminated most high-powered animals. Besides disrupting the positive feedbacks that these animals created with other members of the ecosystem, the depletion of powerful consumers has likely reduced productivity of forests, grasslands, and oceans, and has enabled pathogens and dangerously invasive species of plants and animals to spread to areas where native species have no defenses against them. Insect outbreaks, red tides, chestnut blight, sudden oak death, and freshwater cyanobacterial blooms are all symptoms and consequences of ecological instability arising from the human-caused erosion of controls on population sizes, controls that in undisturbed systems exercise diffuse regulation and maintain economic health.

The problem with environmental degradation is that it does not function effectively as a common enemy. Mobilization of social pressure to confront destruction of what remains is therefore challenging, especially given the short-term benefits of continuing the unsustainable exploitation of nature's resources and services. The dogma of continuing growth is the greatest impediment to solving the ultimate tragedy of the commons: the global degradation of the biosphere.

There seem to be three reasons why this dogma of growth is so entrenched. First, modern economies have become entirely dependent on continued growth. If economic growth improved human lives and brought prosperity in the past, so the argument goes, it should do the same in the future. After all, depleted resources are replaceable and, the argument continues, prosperity brings greater environmental awareness. Moreover, as the economy grows and budgets increase, more resources could be allocated to multiple competing demands, from environmental protection to aiding the poor. Whether such beneficial allocations come to pass depends, of course, on the nature of government and the distribution of power. Too many human societies, including those run by governments that claim to represent the public as a whole, have squandered wealth on wars, unusable weapons, ostentatious projects and lifestyles, tax schemes that favor the rich, and unwieldy bureaucracies.[22]

Perversely, the thesis of this book bolsters the argument for the beneficial consequences of continued growth. The history of life is one of overall economic expansion and lifting of constraints on maximum power. It might therefore seem reasonable to expect that growth can continue indefinitely as long as more limits to power are relaxed and resources that now appear to be in short supply can be substituted with other, less exploited ones.

The problem with this expectation is that, unlike any previous economic system on Earth, we as a species have achieved a monopoly. Some resources such as food are not easily replaced, and others such as freshwater require huge energy outlays to bolster supply through desalination and cleansing. To many of us, the Earth simply cannot sustain further expansion in human numbers and per-capita power.

The second and third reasons for the continued adherence to the dogma of economic growth reflect intellectual limitations. It has proven difficult for economists to come up with credible measurements of environmental degradation that policymakers and the public can readily grasp. Despite objections raised by many economists who are aware of human-caused ecological destruction, I agree with Brian Czech that the classic gross domestic product (GDP), measured as the nationwide or global economic activity by humans per year, is a reliable if incomplete indicator of the human imprint on the biosphere. It is expressed in monetary units per year and is therefore a measure of power of the economy. Czech rightly notes that there is a time lag between GDP and environmental damage, because the latter continues to increase even as GDP decreases when resources are overexploited. Nevertheless, it summarizes economy-wide power in a single number that correlates with resource exploitation and that is already familiar to economists and politicians. Where it fails is in ignoring the accumulated effects of human activity, that is, the increasing size of the physical technosphere.[23]

The third problem is that conventional economics has so far been unable to incorporate a theory of how to construct a healthy economy that does not grow. Stagnation—the preferred word for the steady state—connotes ill health, the specter of recession and economic depression. This diagnosis is so toxic to economists and politicians that the field of economics has been unwilling to contemplate serious reforms to the current capital-intensive economy or to conceive of a less intrusive economic system in which power is distributed more evenly. The fundamental—and mostly unasked—question is this: How can humans reconstruct an economy that will appeal to people, yet function within the limits of Earth's resources? Can the monopoly be undone or its effects on the biosphere made more compatible with the workings of Earth's living system?

Preserving Adaptation and Regulation

With these perils of monopoly in mind, two questions arise. First, have the emergent states of life in nature avoided monopolies and destructive internal conflicts and, if so, how? Second, how might humans avoid such destruction in the future? Does the history of life offer any models for diffusing monopolies and preventing wholesale collapse stemming from internal violence? Is there any way in which adaptation, shared power, and the diffuse regulation that have worked so well in the past can be restored?

The first point to make is that, although prehuman life eventually created conditions favorable to its persistence and proliferation, some early innovations certainly had drastic—and not always beneficial—effects on the status quo. As a power-enhancing innovation in cyanobacteria, oxygen-releasing photosynthesis provided these bacteria with a substantial competitive advantage, but oxygen was toxic to the prevailing anaerobic community, which therefore suffered a contraction to habitats where oxygen did not penetrate. Meanwhile, many organisms adapted to the new world order by evolving aerobic respiration, another power-enhancing innovation. This example illustrates the power of innovation and adaptation to make the environment more congenial to life as a whole. In addition, and importantly, anaerobic life not only persisted in refuges, but continued to perform essential functions in the circular economy that was now dominated by aerobic organisms.

To take one more example, consider the well-established trend for lineages of marine animals originating in shallow water to colonize, and eventually become restricted to, deep offshore habitats. As noted in chapter 7, this happened to brachiopods and sessile crinoids, but the same fate befell numerous other lineages of gastropods, crabs, cidaroid sea urchins, octopod cephalopods, barnacles, and sponges, among many others. The most likely explanation is the evolution in shallow productive habitats of increasingly powerful competitors and predators. This collectively made life untenable for lineages that had evolved there earlier, before their more powerful enemies had arisen. Survivors in shallow water adapted to the novel conditions, but the refugees persisted and thrived on the less productive, deeper parts of the seafloor where the more powerful species could not make a living. Going deep and cold is thus like going back in time. As in the preceding example, adaptation and the availability of refuges were critical for avoiding the kind of creative destruction that labor-saving human innovations have so often brought about.[24]

The problem in much of the human-dominated biosphere is that, besides having eliminated many species, we have removed many refuges and therefore diminished the capacity of many lineages to adapt to new circumstances or maintain the adaptive traits they have in unexploited safe havens. In nature, power even at the top tends to be shared; in today's biosphere, power is in the hands of a single, overwhelmingly potent species. In other words, whereas power in nature is neither absolute nor uncontested, in today's technosphere it is both. Contests for power occur within human society, of course, but from the perspective of the rest of nature, these contests entail only more destruction.

To replace nature's diffuse regulation, humans have had to resort to a more top-down management of our environment through deliberate policies. This transformation from natural to more explicitly human-imposed regulation has at best been only partially effective in preserving the biosphere's resources and services. There is no sign that the trend toward an increasing share of the world's productive resources flowing to humans will cease, let alone reverse. The rate of increase in the human population is slowing, but the population is still increasing, as is the per-capita power of individuals. Restoring the ability of organisms to adapt and reestablishing diffuse ecological regulation are a long way off.

The challenge is not that the ability of individual humans to adapt to changing circumstances is deficient. To the contrary, humans have unprecedented agency, the ability to forge the future, aided by technology and organized coordination. Instead, a primary stumbling block to societal agency in response to rapid physical and cultural change is a relatively rigid body of laws, a sort of societal genome whose foundation in many modern states is a constitution that is designed to be difficult to modify. Like genomes in organisms, constitutions reflect solutions to conditions in the past, and therefore work well as long as the future is like the past. But if the future is very different, strict interpretation of an already inflexible legal code makes societal responses to rapid cultural and technological change extremely difficult. This is equally true for a literal reading of founding religious texts. Strict adherence to a legal framework conceived long ago may lead to consistency but not necessarily to a desirable outcome that confronts current and potential future threats. Gene expression in nature depends on many non-genetic factors and is therefore to a substantial degree flexible in the face of prevailing, often changeable external conditions. Human societies would, I think, do well to emulate that flexibility in treating law as a guide to be flexibly interpreted rather than as an unyielding prescription for problems that the original framers could not have imagined.

Incompatibility among interest groups further contributes to the difficulty of societal adaptation to existing threats. Capitalists and those who benefit from them, for example, owe their short-term success to continued economic growth. Any problems arising from such growth can, according to Julian Simon and other optimists, be solved by brilliant people, whose number will increase as population size continues to rise as the economy expands. This rationale implies that increasing power among the wealthy benefits the poor and perhaps even the biosphere as more resources are allocated to social welfare and environmental protection. The assumption undergirding this reasoning, though never stated explicitly, is that concentration of power engenders positive feedbacks similar to those in the ecosystems of nature. This expectation, however, is unlikely to be realized not only because continued growth is unsustainable (and has been since the early 1980s), but also because of conflicting interests between rich and poor. The poor and powerless in society have aspirations not unlike those of the wealthy, but their capacity to fulfill them is limited by policies such as regressive taxation and the lack of representation in wealth-creating institutions. Bringing this large class of people up to the living standards of the rich is ecologically and economically unfeasible because it would require that our current use of energy would have to rise by a factor of five or more. Within the human economy, therefore, the only rational solution is, as Robert Reich and Thomas Piketty have emphasized, a redistribution of existing resources, to be accomplished by shifts in policy. There would have to be progressive taxation, and inheritance of wealth from one generation to the next would have to be substantially curtailed. There should be strict limits on top individual incomes, and employees should have a say in, and directly benefit from, the profit-making activities of corporations.[25]

These measures would not, and should not, eliminate inequality entirely. As I noted in chapter 7, inequality of power among economies and individuals characterizes all ecosystems throughout the history of life. It is therefore highly unlikely that human society can do away with such differences in power. A primary reason is that incentives for self-improvement, mediated through competition, will be essential to ensure a workable economy and society, much as natural selection—the evolutionary equivalent of incentive—is in nature. In short, inequality can be kept in check by shared power, which in the human economy can be achieved primarily through deliberate policies decided upon by a government in which divergent interest groups are represented.

In so many words, this is the prescription offered by Francis Fukuyama in his analysis and defense of liberalism, a system of government based on rules encoded in laws. In such a system, the goal is to protect individual rights such as privacy and free speech (within limits), to promote equality, and to encourage enterprise that benefits society as a whole. Diverging interest groups are represented, but ideally none of these gains the upper hand of power. Rules of social engagement should be based on rational, objective knowledge gained through the application of the scientific method in which evidence is considered and hypotheses and solutions are modified according to new facts and changing conditions. Liberalism succeeds if self-interested parties compromise and learn to live with each other, not unlike the situations that exist in natural ecosystems.[26]

Fukuyama points out that liberal government and its more restrictive, less democratic antecedents in Europe led to a period of nearly continuous human-economic growth. An unresolved question is whether it can also survive in a world where an economy reaches a form of dynamic equilibrium, one in which collective activity is commensurate with the availability of resources. There is also the question whether a less unequal distribution of income and wealth can be achieved without destroying or strongly curtailing the link between individual ambition and merit. Some critics on the left of the current political spectrum hold that poverty, racism, and other social ills prevent people from taking advantage of educational opportunities, which come with entrance requirements. The solution these critics offer comes down to eliminating selective entry and criteria of merit. To me this solution insults that part of the population that is culturally disadvantaged, because it tacitly assumes that the inability to meet standards is intrinsic. Moreover, it catastrophically diminishes incentives for all those who can, and want to, satisfy the merit-based criteria. If history has anything to tell us, it is that competition and merit are essential to the maintenance of a healthy economic system from which most individuals reap rewards. A world without competition is one of mediocrity. There are so many ways of excelling in human society that merit-based criteria and incentives to meet them enable individuals of many talents to succeed in a competitive economy and society.

Several scholars advocating for a more sustainable human-led biosphere have called for a change in the criteria for what it means to be fulfilled in life. Rather than accumulating material (and therefore energy-intensive) wealth as a symbol of success and status, respect might be conferred by individual or social philanthropy and by a greater value placed on a rich inner life and a deep appreciation of the natural world. Steven Pinker has, to my mind more hopefully than realistically, argued that humanity is already undergoing a process of what he calls dematerialization, where physical objects are increasingly being replaced by resources online. Other means of self-actualization might include enriching experiences, creative arts, rewarding explorations of the mind, and the satisfaction that comes with understanding.[27]

Some of these proposals are probably far-fetched or wildly optimistic, but there is some precedent in nature for changes in the criteria of success, as indicated by novel patterns of natural selection. One example discussed in chapter 7 is the rise of grasses relative to trees in moderately dry climates, whereas grasses respond adaptively to intense herbivory by hiding the growing meristem near or under the ground and letting consumers take the older, less productive parts of the plants. A great benefit of this novel response was that grasses did not have to invest in vast expensive trunks as means to reach sunlight as trees must do. The replacement of passive defenses by active ones and the rise of aggression among predators are other possible examples of changing criteria of success. Interestingly, all these changes in criteria were made possible—and perhaps necessary—by increases in power in top consumers. To what extent, then, could a change in criteria be realized in the human landscape of incentives and norms?

Humans, like all other life-forms, require material resources to live and work. This means quite simply that incentives to accumulate material wealth for greater comfort and status will never go away. We can and must endeavor to rely on more replaceable resources of energy and to recycle more of the waste products we release into the environment, including plastics, metals, and toxic chemicals. Food and water, however, are not easily substitutable. Divorce from material wants is therefore unreachable and probably an undesirable goal. Instead, wealth must be more equitably shared, not just within the human economy but in the biosphere as a whole. In the present state, this can be achieved only by human-crafted policies arrived at through a deliberative process.

Another part of the solution must be an agreed limit on human population size. It is true that, as a population ages, without sufficient replacement, there will not be enough younger people to accomplish essential tasks. In order to make population limits economically feasible and desirable, society must invent incentives to limit the size of families and at the same time lengthen the duration of a working life, especially for young adults. Birth control should become much more widely accepted, and the period of dependence of offspring on their parents should be shortened by making education and training more time-efficient.

That said, I think there is a place for incentives that link status to nonmaterial wealth. In the present state, where economic inequality of income and wealth is still stark, it may be impossible for such a change to take place. In fact, a more equitable distribution of income and wealth may well be a necessary condition for the emergence of criteria that favor a shift toward a more sustainable criterion of status and for the other incentives discussed here to work. I am not qualified to elaborate further on these suggestions, but it is clear to me that the subject of changing criteria of success is in urgent need of serious work by economists, sociologists, and psychologists.

Even the most effective and democratic governments have fallen short in coming to grips with the long-term harm that the human monopoly is causing. I therefore perceive a role for non-political philanthropy—the willingness of the rich and powerful voluntarily to give money and time to projects and causes whose beneficial payoffs are not beholden to the short-term constraints of elections. Environmental protection and conservation, the collection and dissemination of factual information, promotion of the arts, and providing educational opportunities to individuals with culturally inherited disadvantages strike me as realms in which philanthropy can be highly effective and beneficial. In an ideal world, philanthropy should augment rather than replace state policy, and should not exempt people from paying taxes for the common good. Having been a recipient of both state-sponsored education for the blind and philanthropic largesse from foundations supporting my scholarly work, I can attest to the benefits of public as well as private efforts to make a better world. I therefore see the rise of the welfare state and the increasing role of forward-looking organizations without government ties and without a profit motive as hopeful developments.[28]

It is unfashionable to be pessimistic, but I cannot manufacture optimism even as I retain hope. Somehow, humanity will have to invent voluntary limits to its own power and learn to share that power with the rest of the biosphere. This will require actions by many governments. In line with the critical role of adaptation in having kept the biosphere going for billions of years, various policies and solutions must be tried and tested in order to discover which ones work best. Shared power among semiautonomous human groups may be the only way to ensure that this happens without internally generated conflicts getting out of hand and destroying us all.

9

Power, Pattern, and the Emergence of Beauty

E pluribus unum—out of many one—is an apt Latin phrase adorning the currency of the United States. It expresses the hope and expectation that the cacophony of different voices in the populace can be fashioned into a workable, cohesive whole that accommodates the interests of all its members. It is a quest for unity, a belief that something bigger and nobler can emerge from the activities, aspirations, and perspectives of individuals working together.

The same phrase could equally apply to the scientific enterprise, a collaborative effort to uncover unifying principles and regularities from a vast unordered jumble of objects, places, interactions, events, conditions, processes, verifiable facts, and ideas. Science derives its explanatory power not only from observation of phenomena and evaluation of hypotheses but also from the construction of theories that bring these elements together into a coherent framework. Such theories, including the theory of evolution as first laid out by Charles Darwin in 1859, are never the final word, but instead are continually honed and enriched by new findings and new perspectives. Without attempts to build on already firm foundations, science runs the risk of transforming theory into unquestioned dogma, precluding new approaches and thwarting efforts to make sense of what continues to be a world full of mysteries. Science, in other words, is like adaptive evolution; concepts, like organisms, are constantly being tested and, when possible and if necessary, modified to forge a better fit with reality.

It is in this spirit that I have written this book. My intention is not to overturn well-established evolutionary paradigms about phylogeny, the genetic basis of natural selection, and the fundamental connection between building a living body and evolving a lineage. Rather, my aim is to broaden the evolutionary worldview to include a more explicitly economic and historical component, to integrate that narrative with the course of human events, and to probe the nature of history itself. This perspective builds on the labors of countless paleontologists, biologists, and theorists, as well as on my previous books. The present work extends these efforts by assigning a leading role to the pursuit of power in the history of the biosphere and by showing how agency—competition, cooperation, and innovation—has lifted constraints on life and has led to the emergence of new, more powerful states. Power at every level in the hierarchy of life has increased, and has done so at an accelerating pace.

Is there more to history than an increase in power over time? Are there other properties of history that emerge from the condition of being alive? I believe the answer to these questions is a firm yes.

In this final chapter I first propose that innovations and adaptive responses to challenges have become less contingent on circumstances and have therefore become more likely as the role of chance and the yoke of history have receded over time. Feedbacks and outcomes are certainly not inevitable or predetermined, but as adaptations and innovations accumulate, limits on power are eased and many new, highly effective levels of performance become accessible to organisms. This interpretation of history as cumulative and directional clashes with much current thought, which assigns a leading role to chance and to influences from the past on the present. I argue that, although the past cannot be erased, the economic and evolutionary principles of agency and selection that have governed life throughout its history are consistent with, and indeed postdict, the arrow of time toward greater power.

Up to this point in the book I have had little to say about the philosophical implications of the rise in power over the course of life's history, but

I briefly consider three of them here. One is the emergence of freedom, a concept grounded in personal choice of action and thought that has been applied almost exclusively in the human context. I argue that freedom is enabled by greater power, and that it represents unprecedented agency—and responsibility—among living beings.

I then consider how the pursuit of power can be reconciled with the emergence of pleasure, appreciation, love, and a sense of beauty, sentiments that add new dimensions of meaning to an uncaring adaptive foundation. Attractiveness emerges from the advantages of protection against enemies and lays the groundwork for the human pursuit of understanding, an intellectual beauty that enriches and clarifies the extraordinary diversity of present and past life.

Finally, I consider life and its history in the much broader context of the emergence of matter, energy, time, space, history, and power. Life is but one expression of a general tendency for probability, which rules the world of quantum mechanics, to wane in favor of more predictable emergent states.

From Chance to Expectation

Stephen J. Gould famously proposed a thought experiment in which he considered what would happen if we reran the "tape of life" on a planet similar to Earth, where life would have evolved independently. He believed that a replay would yield a history utterly different from the courses of evolution on Earth. In Gould's view, contingency—the dependence of lineages, events, and evolutionary processes on particular, unique prior circumstances—was the predominant agency in the history of life. Natural selection and adaptation have their place in Gould's conception, but they took a back seat to such externally imposed shocks as mass extinctions, continental collisions and breakups, and the particular coincidences that prevailed at the times that major evolutionary lineages originated and diversified. By their unique nature, these prior triggers—all beyond the control of living things—cannot be predicted, and neither could their consequences. Were it not for the unforeseeable mass extinction of the dinosaurs at the close of the Cretaceous, Gould argued, we might never have had the great diversification of Cenozoic mammals and therefore no primate that took over the world. In short, the history of life is a description, a narrative without direction.[1]

Gould's argument in favor of contingency was a necessary retort to the prevailing view that history and evolution are progressive, an idea deeply entrenched in Western thought since at least the Enlightenment of the eighteenth century. Like Gould, I reject notions of progress because of the implication that things have become better over time, a value judgment that only winners and survivors would find compelling. Much evolutionary writing almost reflexively used words such as primitive, lower, advanced, and higher to refer to lineages and traits, rather than more neutral terms like ancestral and derived. Humans were widely considered to be the pinnacle of evolution, and dominant human races considered themselves superior to others whom they labeled as savages, natives, or barbarians.[2]

Philosophers and historians of human affairs have likewise claimed that the consequential events of history are unpredictable and dependent on unique causal triggers, conditions, pathways, and personalities. Isaiah Berlin, Karl Popper, Jacques Monod, and many others rejected the idea that history could be studied as science and that a theory of history can exist, because events and their consequences cannot be replicated. According to these thinkers, no general conclusions can be drawn and, although evolutionary principles can account for what we observe, they cannot serve as the basis of prediction. Most historians—and paleontologists—therefore emphasize particular narratives, in which specific social or ecological conditions and the adaptations (or emotional states) of individuals loom large. In short, students of history view the past as a sequence of particulars, which are worthy of study for their own sake but which in the aggregate do not cohere into a unified account that might reveal trends and patterns.[3]

At one level the critics are right. Historical contingency indisputably reigns at the scale of life's particular, temporary productions such as individuals, lineages, clades, and coherent groups. These productions have specific identities and characteristics that reflect historical legacy as well as adaptation. Likewise, phylogenies comprise unique branching sequences of ancestor-descendant connections. Mutations, mass extinctions, single interactions, battles, treaties, and contracts are also temporary phenomena with distinct beginnings and ends. Within their categories, these productions and events may have features in common and might often have common causes and effects, but their identities bear the stamp of history. The point is that proponents of a decisive role for contingency in history are chiefly concerned with temporary manifestations of life and the events to which these productions are subjected.[4]

Where I believe the critics miss the mark is in failing to consider that life's productions act—that is, do things—in an economic context. Living things are more than creatures with names, traits, and pedigrees; they interact in predictable ways given their power and characteristics, and the outcomes of their activities affect their present and their future. Traits and circumstances are often idiosyncratic and therefore contingent on the past, but actions and outcomes are mediated by power. In other words, the accomplishments of living things—acquisition and defense of resources, establishment of economic relationships, and reproduction—are less contingent on particulars and more dependent on power, no matter how that power might be expressed in adaptation. Outcomes are neither arbitrary nor random. How outcomes come about, how traits and circumstances influence them, and how working organisms are built are fascinating questions whose answers contribute to an understanding of history, but it is action in an economic context that makes history. In history as in life, it matters more what you do than who you are.

In this characterization of living things as agents that interact economically, history is more than a string of names, places, dates, events, and conditions. It is necessary to know what happened to whom when and where, and in the human case to understand motivations and perceptions. But in order to discern the larger themes and trends of history, insofar as they exist, we must go beyond the particulars and delve into performance levels and the feedbacks between agent and environment. This kind of comparative history, founded on economic and evolutionary processes, is most assuredly accessible to scientific inquiry. A search for patterns, trends, and emergent properties can be undertaken without an embrace of historical progress.

If predictable states and trajectories do emerge over the course of time from the actions of contingent organisms living under contingent circumstances, how does this happen? How does chance become expectation if not necessity? This is a question not only for historians, but also for physicists who grapple with the uncertainties of the quantum-mechanical world merging into the material world at the macroscale. Favored states of matter seem to emerge as islands of stability from quantum states. Do the actions of transient beings lead to an emergent economic entity whose structure and dynamics generate a historical directionality?

Some thinkers invoke what they call self-organization to account for the emergence of order from chaos. In this view, when two would-be components come together, they inevitably form a stable emergent structure with new properties. The "self" in self-organization implies that the interaction either releases energy or, if energy is consumed, the source of that energy is predictably available. Self-organization would be involved, for example, in the formation of water from oxygen and hydrogen, and in many similar chemical reactions. It may well have played a decisive role in the origin of life in the confined spaces of warm, metal-rich, alkaline hydrothermal vents. Under these highly specific conditions, the assembly of polymers and even the formation of cells enclosed by membranes could have been essentially inevitable. If this scenario is correct, the inevitable origin of life was highly contingent on the particular environment of alkaline vents. It is notable, however, that although such conditions still exist today, the modern Earth does not seem to be giving rise to the living state from inorganic matter.[5]

When agency by organisms is involved, however, the role of self-organization diminishes. I would argue, for example, that no self-organization should be invoked in the formation of symbioses, the outcomes of competition and predation, or the establishment of collaborative relationships between free-living organisms. These interactions and outcomes result from energy-demanding activity by living things, and are not inevitable. Like organisms themselves, economies are emergent but not self-organized; they are the products of agency and selection, of feedbacks and adaptation.

The actions of transient beings, whose existence reflects a strong historical stamp, yields an evolutionary predictability because of the continuous process of adaptation—a feedback between organism and environment—and the accumulation of adaptive innovations that, for want of a better term, might be called adaptive knowledge. Expected outcomes arise from the repeated actions and adaptive evolution of more contingent bodies.

Natural selection and life's agency are powerful because they reward, and are the products of, success in the biosphere. Agency means that organisms affect their surroundings. If living things improve the environment for themselves and for the organisms on which they depend, subsequent selection and behavior will reinforce those beneficial outcomes by establishing and favoring collaborative economic relationships in addition to competitive ones. Put another way, adaptation by interacting lineages leads to an economy that benefits most of its members most of the time. Moreover, this emerging economy is conducive to innovation and further adaptation insofar as it provides a stable and accessible pool of resources under permissive enabling conditions. Such an economic system is reasonably resistant to external shocks and, by virtue of the adaptive potential of its members, capable of recovering from disruptions, at least until some part of that economy achieves monopolistic power.[6]

This process must eventually come to an end, either when constraints on resources prevent further increases in power or when a state of monopoly is reached. The fact that the power of the biosphere has trended upward throughout the history of life is testimony to the power of serial and accumulated innovations to relax the limitations imposed by resources.

These innovations have enabled the rate of turnover of existing resources to increase and have made additional resources available that previously had not been exploited. The time and place of innovation are to some degree unpredictable, especially during the early stages of evolution, but as adaptive innovations spread and accumulate, they and their consequences render the long-term trajectories of history, including the imposition of hard limitations, more predictable.

Metabolic evolution lifted the constraints of the hydrothermal environment and enabled life to escape to a much broader range of habitats and to change those environments in its favor. With these biochemical innovations—the Krebs cycle, photosynthesis, nitrogen fixation, and more—life became less dependent on particular conditions, and its persistence was no longer as contingent on specific circumstances. Constraints on power were still severe, but as additional biochemical and anatomical innovations accumulated in many lineages, and as ecological relationships were established, limits on power continued to be eased.

Innovations kept coming. One of the most crucial for adaptation was the evolution of sexual reproduction, in which the common interest of independently assorting genes is tested and rewarded when recombination after meiosis places genes into new genetic environments in which they must function. Genes become interdependent rather than autonomous units and must work together in a workable genome that remains flexible enough to incorporate changes that affect the function of the entire organism. Sex evolved as the common good on the genetic level.[7]

The conditions under which sex first evolved are not well circumscribed, but many subsequent innovations must first have arisen under very specific conditions even if later versions evolving in other clades could arise in a greater range of circumstances. There was a first time when mineralized skeletons arose, likely under carbonate-rich conditions in shallow marine waters during the Late Ediacaran. Lineages in which mineralization evolved later and independently lived in various environments, including terrestrial ones. Likewise, there was a first time when one or another lineage evolved parasitism, internal fertilization, eusocial organization, endothermy, venom, image-forming eyes, appendages, jaws, a bivalved shell, jet propulsion, active flight, plankton-feeding, electric shocking, agriculture and domestication, a vascular conducting system, leaves, roots, flowers, fleshy fruits, basal plant growth, and insectivory by plants. Each of these noteworthy, energy-demanding, and energy-enhancing innovations initially evolved under very particular circumstances, but all of them evolved again and again independently as the conditions under which they were beneficial became more common and widespread. Only one or two lineages might have evolved the predisposition necessary for the origin of the earliest iteration of a new structure or capability, but with the accumulation of innovations in other lineages, more and more lineages in different environments reached the point where similar innovations could arise and provide adaptive benefits. In other words, an increasing number of lineages and environments became receptive to energy-enhancing innovations. Endothermy may have begun in small animals in warm climates on land, but it subsequently evolved in or spread to the sea. Basal plant growth arose on land in response to intense herbivory, but also evolved separately and later in the sea as hungry, secondarily marine grazing mammals became important selective agents. In short, the initially large role of chance gave way to a more robust, predictable, and permissive environment and an increasing number of adaptively receptive lineages.[8]

The sequence from contingent first appearance to expected subsequent origins under a much wider range of conditions recalls the origins of humans in Africa and the later human ascent to modernity (see chapter 8). Here, too, the circumstances of origin, although perhaps not as exceptional as those under which life originated, are quite specific and certainly not universal. The evolution of our ultrasocial characteristics can therefore be considered highly dependent on conditions present on the African continent during the Early Pleistocene. Once it evolved, however, ultrasociality set in motion more predictable trends and feedbacks in many places.

To be sure, there are innovations—and human inventions—that do appear to be unique. The formation of the eukaryotic cell is one, although it does represent an instance of the very widespread phenomenon of symbiosis. Nitrogen fixation, oxygen-producing photosynthesis, the nervous system of animals, and human language may be others. Most of these purportedly one-of-a-kind innovations (except for language) occurred very early in the history of life. This means either that life in those ancient times was considerably more contingent on particular circumstances than it has become since, or that the mists of time have obscured multiple acquisitions, perhaps in closely related lineages. Both possibilities seem plausible to me. Either way, the early rare breakthroughs clearly benefited the organisms in which they occurred because they lifted constraints on power. Their persistence, in other words, is no accident even if their first or only iteration was.[9]

These examples show that innovation is a two-step process. First, a new state must arise. Whether through mutation, chromosomal rearrangement, gene duplication, or behavioral novelty, this first step of innovation is significantly influenced by chance. It is probably a stretch to claim that these initial events, including mutations, are random, as is often asserted. Because of the three-dimensional structures of nucleic acids and especially proteins, some mutations or rearrangements are much more likely than others. Moreover, some modifications, such as a change in gene function or a tweak in the sequence of developmental steps, are more likely to arise in some lineages than in others simply because the genetic architecture and anatomical structure of the developing body are especially conducive to modification. A veined leaf, for example, is more likely to arise in a plant with a vascular conducting system than in a seaweed in which a leaf-like thallus exists but a vascular system is absent. Most innovations arise as modifications of existing structures and relationships, meaning that predisposition is an important circumstance in the origin of evolutionary innovations.[10]

The second step is acceptance. This means that the innovation must be useful to the organisms in which it arises, so that it can evolve by natural selection to benefit the next generation. It also means that there must be opportunities for the innovation to spread. Many animals and plants on isolated oceanic islands, for example, have evolved characteristics found nowhere else, but they and their innovations go nowhere as long as the lineages are confined to their place of origin. Examples of such unusual innovations include grass-eating by flightless parrots in New Zealand, the tree habit of otherwise mostly herbaceous Asteraceae on the Galapagos and Mascarene Islands, herbivory by a secondarily marine iguana in the Galapagos, and an additional jaw joint in snakes on Mauritius and nearby islands. These new states work well where they evolved, and in some cases might have done so had they arisen in or spread to more competitively rigorous continents, but isolation together with the lower performance standards on remote islands doomed these adaptive departures from the norm to parochial stardom.[11]

Human inventions go through a similar two-step process. Many very clever people invent potentially useful machines or techniques, but these inventions become important and undergo further development only when they spread and become socially accepted and adopted. As they become widely used, the inventions are increasingly independent of the original conditions under which they were conceived. Here again, however, as in the origin and evolution of biological innovations, some conditions are socially much more conducive to tinkering and exploration than others, implying that the origin of new discoveries is not as contingent as some might think.[12]

Unpredictable events always remain, of course. They are unpredictable precisely because organisms are not adapted to them. If events are destructive, as they usually are, some organisms might be better predisposed to resist because the disruptions somewhat resemble minor and more frequent circumstances to which organisms are well adapted. During mass extinctions, for example, when primary productivity plummets, organisms living under chronic food shortages might be better equipped to withstand the interruption than those dependent on a predictably high food supply. If levels of carbon dioxide spike during such crises, organisms capable of tolerating high levels of this gas in their tissues or resistant to the inevitably greater acidification of seawater would be more apt to weather the crisis than more sensitive lineages. These effects of rare extinction events might be long-lasting, with the result that subsequently evolving lineages retain their tolerance or resistance and their predisposition to survive another similar crisis. Over time, even rare and unpredictable events can thus become less disruptive.[13]

Not all rare events are bad. Gigantic volcanic eruptions that release vast quantities of nutrients such as phosphorus and iron into the biosphere can stimulate primary production, especially if eruptions are submarine and do not poison the atmosphere or limit the penetration of sunlight. Such submarine flood-basalt eruptions are known from many periods in the geological record, especially during the Mid- and Late Cretaceous. These windfalls can benefit organisms only if the nutrients can be quickly captured and retained.[14]

What, then, would happen if the tape of life were rerun? I contend that the broad outlines of history would be the same regardless of where life evolves. Many adaptive solutions to competitive challenges would be similar to those that evolved on Earth. How far any of these trends and solutions could be realized depends on such factors as temperature, habitat size, chemical composition, gravity, and the amount of heat in a planet's interior for generating tectonic activity and the movement of materials to and from the planet's core. Even on our own planet, life has followed parallel trajectories regardless of time and place. Likewise, the human economy has followed similar sequences of change in different parts of the world. Chance looms largest early in history and in the realm of who and what the particulars are, whereas trends and economic outcomes emerge as expectations from adaptive processes.

An Evolutionary Perspective on Freedom

It may not have escaped readers that there is an intimate connection between power and freedom. By eliminating or overcoming constraints on agency, greater power expands the range of possible actions by individuals or coherent groups. Although the notion of freedom is typically considered only in the human context, it applies to all life-forms and can therefore be approached from an evolutionary and historical point of view. Freedom—the availability of choice of action and, in the case of humans, belief—is a manifestation of enhanced power acquired and wielded by living beings. Constraints on power, whether they are imposed by physical limitations such as temperature and the lack of oxygen or by interference from living enemies, place bounds on what individuals or groups can do and when they do it. Freedom of action is severely circumscribed in animals that are always under threat of predation and in those that perforce occupy confined spaces. Even when constraints are relaxed, individuals with little power cannot immediately take advantage of the greater freedom available to them because they lack the metabolic power to gain access to new opportunities. In the long run, however, potential can be transformed to realization through a combination of power-enhancing innovations, selection for greater power, and the resulting greater agency, the ability to modify surroundings and to locate environments where conditions for resource acquisition and defense are better. In short, power and freedom evolve together.

There are two important points to make about freedom and power. First, constraints are never entirely eliminated. In fact, competition and the requirement for defense are essential conditions for distinguishing between adaptive innovations and novelties that are harmful or that in human terms would be considered frivolous. Freedom of action is not an unambiguous good; it is a benefit only if the choices made offer a payoff to the individual or group. Just as power is usually subject to some upper limit set by physiology or the environment, freedom is never absolute; it is constricted by the actions of others and, in the human case, by social norms.

Second, freedom is like power in that it is not evenly distributed. Powerful beings simply have more freedom, more choice, and more influence over others than those with less power. In human society, powerful members also bear more responsibility—the idea that their choices have consequences not only for themselves but for other members of society and the biosphere as a whole. This responsibility creates its own limitations on freedom. Greater power thus enables a greater range of choice but also imposes limits on those choices; it shifts constraints from collective limitations imposed from outside to the actions of the most powerful members. As the difference between the most and the least powerful members of a system widens, as I argue it has done over the course of history, the fate of systems is increasingly in the hands of members with the greatest power and freedom.

Some human societies have created political systems that expand individual freedoms and that extend those freedoms to a broader slice of the population. Through representation, these democratic systems offer individuals some say in how they—the powerful and the powerless alike—are governed. Democracies do not erase inequalities in power or freedom, but they have broadened opportunity and choice. They are always under threat of authoritarian actors—individuals, economic entities, and the state—which monopolize power and restrict freedoms for all but a small, privileged elite. Such tendencies, too, have their analogs in nature. Lineages that at one time held positions of power in ecosystems are usurped by new elites and are subjected to increased limitations imposed by the ascendant hegemony. Power and freedom thus are not permanent states, but are subject to the economic and selective instabilities that are inevitable given the perpetual pursuit of power.

This view of freedom from an evolutionary perspective makes it clear that freedom as it is understood in human political and philosophical terms is a recent development in the history of life, one that is intimately connected with the increasingly rapid trend toward increasing power. It also clarifies how access to freedom can change according to the evolving power relations within systems, and why monopoly is bad for the realization of freedom.

Power and Esthetics

Many readers may find my emphasis on power as a unifying principle depressing. Is there really nothing more to life than power? What about beauty, morality, love, and the other sentiments that give life meaning? Are they all to be stripped of their magic in favor of a concept that we often associate with the opposite of these sentiments?

There are two reasons why I find the approach both uplifting and intensely enriching. First, these sentiments would not exist without the pursuit of power. Second, the unity of ideas that the exploration of power offers is itself intellectually deeply satisfying because it explains a vast range of phenomena that at first glance seem unconnected.

As a scientist who has extensively studied shells and who adores the beauty and diversity of nature, I have always considered esthetics to be a prime motivator in my life. I derive pleasure and inspiration from the mathematical regularity of shells, the shapes and textures of leaves, the soundscape of birds and insects singing in forests and fields, the fragrance of flowers, the feeling of warm air rising from a tropical sea at night, the roar of surf crashing on the algal ridge at the edge of the reef on Pacific islands, and above all the discovery of something unexpected and strange. I glory in nature's richness despite the fact that the visual dimensions of beauty are inaccessible to me.

Where do such sentiments come from, and how might they have evolved not only in humans but in many other life-forms as well? The answer is simple: beauty works. The perception of things as beautiful or attractive is no mere evolutionary by-product of some other vital function. It stems from the essential role that sensation and the interpretation of signals play in the lives of organisms with agency. The evolution of mate choice—an adaptive response by many independent lineages to the threat of predation and interference from other organisms—ushered in an era of communication at a distance, where potential mates create and receive signals that others find attractive. The effectiveness of these signals—that is, their attractiveness—affects success in mating and reproduction. Richard Prum has suggested that some courting displays are so power-demanding and so elaborate, or that one party so completely dominates the other, that the costs interfere with reproduction and are therefore unacceptably high. I would counter that the signals work, no matter how expensive and excessive they may appear to us. Signals have many other functions, of course, including detection of food and danger, warning enemies of retaliation, and communicating false or misleading information. It is in any case not surprising that attractiveness is so important in animal and plant communication, because signals come to be associated with mating, food, and protection. Importantly, greater mobility—one aspect of greater power—enables and demands signaling at a greater distance, which in turn means that the signals can become increasingly attractive. In short, I see no conflict between power and the emergence of attractiveness. In fact, beauty as an expression of attractiveness should evolve with power.[15]

Can this happy symbiosis between power and esthetics be sustained in a regime of monopolistic dominance? Human civilization has been wonderful in many ways, not least in having produced stunning works of art, architecture, literature, music, and science. It has unfortunately also produced repulsive ugliness and misery, pollution and poverty, degradation and disorder. There has been progress here and there in reducing air pollution and in conserving bits of the natural world, but I worry that we are collectively losing some of the esthetic sensibilities that nature has bequeathed to us. As an ultrasocial species that has eliminated many biological threats, we mostly interact with other humans. With the majority of the human population now living in cities, many people have little or no exposure to the natural world. Even when they do visit forests or seashores, or when they encounter the products of evolution in zoos or museums, people tend to go in groups, and remain more attuned to social interactions than to the creatures and objects on display. As a matter of personal taste, I derive little esthetic pleasure from most popular music or from atonal, often explicitly unlyrical, so-called serious music. To my ears these twentieth-century and later forms convey the chaos and ugliness of a noisy urban soundscape. The mistrust of intellectuals and the disdain for scientific inquiry are distressingly common themes throughout human history, from the burning of the library at Alexandria to Stalin's purges in the Soviet Union, the Holocaust in Germany, and the rejection of scientific findings about evolution and climate change in our own day.

Our cultural adaptations are focused primarily on our social selves and on our technology. Even so, I celebrate the profound improvements in understanding the natural world. This is in fact my second reason for finding satisfaction rather than emptiness in the study of power and its history. The productions and phenomena of living and past nature take on more meaning when, in addition to studying and appreciating them separately, we can explain and understand them with unifying concepts founded on verifiable evidence and ideas. There is surely no grand plan in the universe or here on Earth, but we have the power to discover the principles and patterns that govern life and its history, and to apply what we know for the mutual benefit of nature and ourselves.

From Uncertainty to Emergence

This book is the story of life and its history. As I see it, life is a state of matter and energy that emerges from interactions among nonliving components and that has the unique properties of agency, natural selection, and adaptive evolution. In the present chapter I have argued that life and its history have become more predictable over the course of time because of the establishment and evolution of economies in which feedbacks and collaborative arrangements mediated by power among living things play decisive roles. As the scale of inclusion and interaction increased over time, chance and contingency receded while predictability and greater certainty emerged.

Is such a trajectory unique to life and its history? I suggest that the answer to this question is no. Life represents a particular form of emergence, but it is not the only expression of that property. The theory of quantum mechanics, developed by physicists in the twentieth century, holds that the components of emergent matter and energy can be described in probabilistic terms of either discrete particles or waves but not both, and that both the position and velocity of a particle can be measured but not at the same time. The foundation of what we as living beings perceive as real therefore rests on probability. It is only when the elementary waves or particles interact that matter, energy, space, and time emerge in the macroscopic domain. The role of probability diminishes but does not disappear. States of matter form transient bodies whose establishment and fates are still subject to chance events. At larger scales of space and longer scales of time, however, the influence of chance wanes in favor of longer-lasting interactions and structures. Emergence would therefore be an expression of predictability and the canceling of probabilistic states. History itself is an emergent process and sequence of interactions with defined consequences, a phenomenon that takes on additional emergent meaning and predictability when it involves the economics of life.

If this conception has any merit, it places life and its time course in a much broader context of all matter and energy, from probabilistic quanta to the formation of economies and galaxies. Power as a determinant of agency and structure emerges as a unifying property of systems, which in the case of life affects the fates of all units in the hierarchy of the biosphere.

NOTES

Preface

1. For considerations of power in the human sphere, see the fine works by Galbraith (1983), van Dijk (1989), Cheng et al. (2013), and Bartulevicius et al. (2020). The essay by van Dijk is particularly concerned with the means of communications by which power is exercised and received, and not so much with the properties that imbue some individuals and groups with more power. All the writers cited, and many more besides, treat power in an exclusively human context, especially in terms of social interactions. Prehuman and nonhuman aspects of power are not only ignored but most likely would not be considered as legitimate expressions of the phenomenon that I argue is a characteristic and fundamental property of all life.

2. See Picketty (2020).

3. See the important works of McShea and Brandon (2010), McShea (2016), and especially Brandon and McShea (2020). The following paragraph in the text is also based on the work of these authors. I return to a consideration of their approach in chapter 3.

4. For major transformations in the history of life, see Maynard Smith and Szathmary (1995) and Lane (2009, 2015). For the transition from molecular to morphological evolution and its consequences, see Butterfield (2000) and Knoll and Bambach (2000). Like me, Coen (2012) attaches great importance to competition, natural selection, and feedbacks; but unlike me he regards the outcome of competition as either survival or death. He misses the important point that losers in competition not only often survive, but help to expand and stabilize ecosystems. I discuss this issue further in chapter 1.

5. Books on other aspects of evolution include the following. For general, more gene-based accounts, see Coyne (2009) and Leigh and Ziegler (2019). The origin of life and the early chemical history of life is beautifully covered in de Duve (2005) and especially Lane (2009, 2015). A lucid treatment of evolution and development is that of Carroll (2005).

6. See Syvitski et al. (2020) for a thorough documentation of the Anthropocene.

Chapter 1: The Nature of Life and Power

1. My emphasis on agency reflects an essentially economic perspective on adaptive evolution, by which I mean the idea that organisms do things as competitors with other organisms (see also Vermeij, 1999, 2004b). Corning (2005) and Kauffman (2008) likewise championed a more holistic economic view of evolution, and like me found the gene-centric consensus emanating from Neo-Darwinism and the Modern Synthesis (as in E. Mayr, 1963) insufficient and

too reductive. Corning and Kauffman prefer the words "purpose," "purposeful," and "teleonomy" to describe activity, in the full realization that those words imply goal-setting. Turner (2007) likewise uses language of this kind to distinguish behavior and physiology from natural selection, which has long been recognized as unable to predict future circumstances. Although I am highly sympathetic to the views of these writers, I prefer the term "agency" because it is descriptive and does not imply, nor does it exclude, a goal.

2. For a discussion of *psuche* in the context of the history of science see Strevens (2020), who faults Aristotle for not using this somewhat vague concept as the basis of a modern scientific theory. To me, even giving the concept (and readily observed phenomenon) a name is a step in the right direction.

3. See Martens et al. (2015).

4. Ibid.

5. For the function of pulvini and leaf movements, see Dean and Smith (1978) and Rodrigues and Machado (2007). For a comprehensive study of the stomates and guard-cell movements, see J. W. Clark et al. (2022). For forces generated by the leaves of rosette plants, see the very clever paper by Sicangcon et al. (2022).

6. This perspective is similar to that of Oudman and Piersma (2018), who use the Dutch word gedrag (behavior) instead of agency. Like me, they accept natural selection as a mechanism for adaptive evolution, but they point to a large number of ways in which organisms and the environment modify and often complicate direct genetic influence. From the perspective of a highly gene-centric population geneticist, Michael Lynch (2007) argued that natural selection plays a distinctly subservient role to such processes as mutation, recombination, and genetic drift. In Lynch's view, the architecture and processes of the genome reflect non-adaptive phenomena. Although this conclusion can be contested, as it was by Lane and Martin (2010) who pointed to various adaptive processes molding the origin of eukaryotes, it perversely (and surely unintentionally) supports my contention that natural selection, agency, and adaptation derive from interactions of whole organisms. Daniel Janzen (1985) also made the point that organisms are well fitted to their surroundings not just by virtue of their genes, but also because they have some capacity to choose the environment they occupy.

7. Nowak and Ohtsuki (2008) distinguish between prevolution, in which a form of selection causes longer polymers to survive better than shorter ones, and evolution, which involves replication of those longer polymers. They argue that selective processes not involving replication precede genes-based natural selection. This seems entirely reasonable given that many chemical reactions are selective without involving replication. It can thus be said that natural selection is a particularly potent process unique to the living world. Chaisson (2005) provides a thorough account of non-adaptive evolution of the universe, and Arthur (2009) does the same for human technology as one manifestation of cultural evolution. In the latter case, evolution can be said to be adaptive because human engineers act as sentient agents who design workable machines and other technologies with deliberate intentions and goals.

8. See Vermeij (2015b) and Karban et al. (2019).

9. See Kauffman (1993, 2000), Corning (2005), and S. M. Carroll (2010).

10. See Tanner and Beevers (2001).

11. Excellent discussions of the mechanics of filter-feeding appear in Rubenstein and Koehl (1977) and LaBarbera (1983).

12. Dressaire et al. (2016).

13. For background, see McCutchen (1977) on samaras; Norberg (1973) for wind-dispersing winged seeds; Niklas (1983) for the mechanics of wind pollination; Stevenson et al. (2015) for the origins of plants dispersing fruits and seeds in the wind; and Timerman and Barrett (2021) for a comprehensive account of wind pollination.

14. See Francis (1991).

15. This topic is thoroughly dealt with by Denny (1988), Koehl and Alberte (1988), and Stewart and Carpenter (2003).

16. Yamazaki (2011).

17. In recent years a theoretical framework called the metabolic theory of ecology has been developed in which metabolic rate, taken as a constant, is related to body mass. For animals, metabolic rate is said to scale as the three-quarters power of body mass (JB West et al., 1997; Brown et al., 2004). Glazier (2015) has criticized this framework, noting that metabolic rate changes dramatically within the lifetime of an individual, even one whose mass remains constant, and that metabolic rate is a highly regulated physiological function. In any case, scaling laws tend to suppress biologically important variations, which different lineages have exploited as they pursue radically different ways of life.

18. See the fine papers by Marden and Allen (2002), Cloyed et al. (2021), Kott et al. (2021), and Meyer-Vernet and Rospars (2017).

19. Data are summarized in Albers and Jarrell (2015).

20. Marden and Chai (1991).

21. Summarized in Alexander (1977).

22. See the excellent reviews by Fratzl and Barth (2009) and Ilton et al. (2018).

23. See the outstanding paper by Hofhuis et al. (2016), who investigate every aspect of silicle explosion in Cardamine.

24. For a summary, see Skotheim and Mahadevan (2005).

25. A fine summary is provided by Galstyan and Hay (2018).

26. Burrows (2014).

27. Among other papers see Patek et al. (2004), Patek and Caldwell (2005), and Devries et al. (2012).

28. This spectacular case is documented by Longo et al. (2021).

29. Experiments by DeLong (2008) show that powerful small animals always win when competing with weaker rivals. DeLong therefore advocates for what he calls the maximum power principle, which is similar to the idea expressed in this book.

30. Vermeij (1982a).

31. Vermeij (1982b).

32. See the outstanding book by Leigh (1999).

33. Van Valkenburgh (1991), Van Valkenburgh and Hertel (1993).

34. Brodie et al. (2002), Dietl (2003, 2004).

35. Vermeij (1982b, 1999, 2004b, 2013a). Dawkins and Krebs (1979) famously proposed the so-called life-dinner principle: When a predator interacts with the prey, the stakes are higher

for the prey than for the predator, because the prey loses its life whereas the predator only loses a meal. This principle ignores the fact that many prey individuals survive encounters with predators, as well as the reality that predators must contend with their own competitors. Van Valen (1976) proposed the Red Queen hypothesis, in which evolving lineages keep "running in place" to keep up with their continuously evolving competitors. We (Vermeij and Roopnarine, 2013) criticized the Red Queen metaphor on two grounds. First, it was invoked by Van Valen to explain what he perceived as a stochastically constant rate of extinction of lineages, a pattern that has not held up to scrutiny. Moreover, most extinctions are not directly caused by other organisms, but rather are due to external agencies such as celestial impacts. Second, evolution is not continuous but episodic, because constraints from many quarters prevent adaptive evolution even when the latter would be advantageous. In spite of these objections, both Van Valen's Red Queen and Dawkins and Krebs's life-dinner principle reflect the primary importance of organisms in adaptive evolution.

36. Itescu et al. (2017).

37. See Birkeland and Dayton (2005), Fenberg and Roy (2008), and Darimont et al. (2009, 2015).

38. Vermeij (2012b).

39. For an excellent review of sexual selection see Clutton-Brock (2007).

40. Emlen (2008), Somjee et al. (2018).

41. Zahavi (1975).

42. Prum (2017).

Chapter 2: What Limits Power?

1. Although feedbacks have been widely recognized as crucial in ecology and evolution, a simplistic one-way cause-and-effect outlook still pervades much current scientific work. Countless scientific papers still carry words like "driver" and "driving" in their titles, implying that some process or circumstance is responsible for some phenomenon being investigated. Rarely if ever is it really established that the purported cause "drives" the alleged consequence. A causal connection is indeed often evident, but whether the link is the one-way ratchet that terms like "driving" imply is rarely established. In a similar vein, many titles promise that something "explains" something else when in fact all that is demonstrated is a description or correlation rather than a cause-and-effect or feedback-related explanation. These criticisms are not just a question of semantics. They reveal a tendency to over-interpret evidence and to confuse pattern with process, description with explanation, and simple cause with two-way feedback.

2. See Rosing et al. (2006).

3. See A. R. Smith (1973) and especially Leigh (1999).

4. These interactions were first discovered by Judy Lang (1971, 1973), who made spectacular observations on corals in Jamaica. See also fine papers by Wellington (1980) and Bradbury and Young (1982).

5. See Rosing et al. (2006).

6. Data in Bar-On et al. (2018).

7. See Cyr and Pace (1993) and Shurin et al. (2006).

8. See Birkeland (1989) and Vermeij (2011b).

9. Excellent reviews in D. C. Smith and Bernays (1991) and Childress and Girguis (2011).

10. Although some aspects of this relationship were known earlier, the most thorough documentation supported by careful experimental work is by van der Heide et al. (2012).

11. A good summary of very extensive work is in Strullu-Derrien et al. (2018).

12. Important papers are by Kiers and Denison (2008), Kiers et al. (2011), Kiers and West (2015), and Epihov et al. (2017, 2021).

13. See Kanso et al. (2021).

14. See McLean (1974, 1983), Vermeij (1977), and Bertness (1981).

15. See Mann (1973).

16. See Osborne and Sack (2012).

17. See Vander Wall (2001) and Vander Wall and Jenkins (2003).

18. This is well documented by Ziebis et al. (1996).

19. Reviewed in Vermeij (2017b).

20. See Rubenstein and Koehl (1977) and LaBarbera (1983) for excellent discussions of filter-feeding. The habitat expansion made possible by active filter-feeding was discussed by Vermeij (1987).

21. See Goldbogen et al. (2007, 2019).

22. See Vermeij (2016a) and Ferrón (2017).

23. Good documentation in Roman and McCarthy (2010), Roman et al. (2014), Savoca and Nevitt (2014), and Lavery et al. (2014). The argument that increased ocean productivity alone, sparked by increasing ocean-water turnover as the result of the onset of high-latitude glaciation during the Pleistocene, was the factor most responsible for gigantism in baleen whales (Pyenson and Vermeij, 2016; Slater et al., 2017) now seems incomplete to me.

24. See Vermeij and Lindberg (2000), Vermeij (2010), Doughty (2017), and Laakso et al. (2020).

25. For summaries see Vermeij (2010, 2017b).

26. For the general case of the stimulatory effect of burrowers, see Thayer (1979, 1983) and Tarhan et al. (2021); for work on fiddler crabs see Bertness (1985).

27. See Abraham et al. (2022).

28. For thorough discussions of temperature see Denny (1993), Gillooly et al. (2001), and Vermeij (2003).

29. Data are from Dreisig (1981).

30. See Kozlov et al. (2015). This study refers to damage by insects but did not include herbivory by ptarmigans (*Lagopus* spp.) or the reindeer *Rangifer tarandus*. It is therefore unclear whether the difference between forest and tundra would hold up if these metabolically more active vertebrates were included.

31. Poore and coauthors (2012) suggested that grazing intensity of marine plants is highly variable and varies little with latitude but Heck and colleagues (2021) and Vanderklift and others (2021) shows that Poore's group seriously underestimated grazing in tropical seagrass meadows. The study by Poore's group was based on comparisons between enclosures that excluded herbivores and nearby areas that were exposed to herbivores. They therefore excluded observational studies or extremely intense herbivory by fishes on reefs where human exploitation

had not seriously depleted fish abundances (see Welsh and Bellwood, 2014). Similarly, a study by Whalen and colleagues (2020) purported to show that consumption of tethered squid pieces by small and medium-sized fishes showed a decline in intensity from the outer tropics to the warmer inner tropics near the equator. Besides a dearth of data for near-equatorial sites, this study may reflect a modern-day reduction in predation of these relatively small fishes by larger vertebrates. Large pelagic predators range widely, but they are especially prominent in warm tropical waters. For compelling accounts of the former importance of large marine consumers and the numerous consequences of the loss of these animals, see Jackson (1997) and Estes et al. (2011, 2016).

32. See Ashton et al. (2022). A smaller but equally impressive study was carried out by Freestone et al. (2021) on the Pacific coast of the Americas. That study's results were similar to those of the Ashton group.

33. For fish endothermy see Block et al. (1993), Little et al. (2010), and Wegner et al. (2015). For bird and mammal endothermy see Ruben (1995) and Lovegrove (2017).

34. For large sea turtles see Paladino et al. (1990) and Sato (2014). Excellent discussions of dinosaur physiology are in Clarke (2013) and Grady et al. (2014).

35. See Seymour and Schultze (1996) and Seymour et al. (2003).

36. Excellent discussions in Verberk et al. (2011) and Rubalcaba et al. (2020).

37. A summary in Vermeij (1993).

38. For a very thorough discussion of the physics of air and water see Denny (1993).

39. See Iosilevskii and Weihs (2008), A. M. Wilson et al. (2013). Throughout this book I am concerned with absolute speeds, measured either as distance covered per unit time or in some cases as number of body lengths traveled per unit time. It is, after all, absolute values that determine an individual's success. For the same reason, I prefer absolute measures of power instead of the more commonly used metric of power per unit mass. The latter way of expressing power may facilitate comparisons of power at the cellular level, but life or death depends on the power of the whole organism.

40. See the very interesting paper by Zona and Christenhusz (2015).

41. For swifts staying in the air for more than two hundred days see Liechti et al. (2015); for insects and their weeks-long migration between Europe and Africa see Hu et al. (2016).

42. See the enlightening paper by Strathmann (1990); also Vermeij and Grosberg (2010).

Chapter 3: Patterns of Power

1. Piketty (2014, 2020).

2. See, for example, Stanley (1973).

3. McShea and Brandon (2010).

4. See Vermeij (2009). Both the phylogenetic and economic interpretations of species and diversity differ from the sociological concept of diversity, which emphasizes categories such as race, gender, sexual orientation, disability, and whatever other criterion of difference has raised the consciousness of society.

5. For the connection between sexual selection and diversification, see West-Eberhard (1983) and Bush et al. (2016). The rise of terrestrial diversity during and after the Cretaceous,

attributable to the diversification of plants and arthropods with intense mate competition, was discussed by Vermeij and Grosberg (2010).

6. See Leigh et al. (2014). It is possible that some of the native South American mammals, especially predators, became extinct at 3.3 million years ago not as the immediate consequence of the arrival of North American carnivores but because of disruptions resulting from the impact of a meteorite in the Pampas (Carrillo et al., 2020). Even if this were so, the new predators rapidly replaced the old guard.

7. For the general case see Benton (1983). For symbiotic bivalves see Vermeij (2013b), and for the large planktivores see Stiefel (2021). Human-historical examples come from Chua (2007) and J. Darwin (2007).

8. For a marvelous book on the history of Christianity see MacCulloch (2010).

Chapter 4: Evolution of Size

1. Payne et al. (2009).

2. Lane and Martin (2010).

3. For the evolution of multicellularity see Grosberg and Strathmann (2007), Butterfield (2009), and Niklas and Mewman (2013). Late Ediacaran sponges were recognized in the study of pumping by Suarez and Leys (2022).

4. For early dates see E. C. Turner (2021). For Late Cryogenian origins see Wallace et al. (2014), Yin et al. (2015), and Brocks et al. (2017). Antcliffe et al. (2014) and especially Bobrovskiy et al. (2021) advocate for a still later origin.

5. For interpretation of rangeomorphs as stem-anthozoans, see Butterfield (2022); very large specimens were described by Ghisalberti et al. (2014) and Xiao (2014).

6. Heim et al. (2015).

7. For the oldest algae see Butterfield (2000) and Xiao and Dong (2006). For larger algae see summaries in Vermeij (2012b, 2017a). Large *Avrainvillea* were described by Littler et al. (2005).

8. For the Silurian see Retallack (2015). A summary of maximum tree heights is given in Vermeij (2016a).

9. For living trees see Sillett et al. (2010). Osborne and Beerling (2002) discussed tall Cretaceous trees.

10. See Retallack (2015) for the Bloomsburg fungus, and Retallack and Landing (2014) for the gigantic *Prototaxites*. Fungal affinities of these fossils were confirmed by Boyce et al. (2007) and Edwards and Axe (2012). Honegger et al. (2013) interpret smaller Early Devonian fungi as lichens and point out that the absence of symbionts in the large fungi does not necessarily exclude the status of these giants as lichens. Nelsen et al. (2020) argue that lichenized fungi arose after the origin of vascular plants.

11. Guerrero et al. (1986), Perleman et al. (2008).

12. For early eukaryotic predation see Kurland et al. (2006). The repeated evolution of predation by eukaryotes employing phagocytosis is discussed by Mills (2020). Predation by perforation is documented by Porter (2016) for amoebae and by Bicknell and Paterson (2018); see also the earlier report by Bengtson and Zhao (1992).

13. The hypothesis that early animals consumed dissolved organic matter was proposed by Laflamme et al. (2009) and supported by Budd and Jensen (2017). This mode of feeding in living sponges is discussed by de Goeij et al. (2008, 2013) and Bart et al. (2021). Cavalier-Smith (2017) has questioned this hypothesis and instead proposed that these early organisms fed piecemeal on prokaryotes, as sponges also do today.

14. Rahman et al. (2015).

15. For *Anomalocaris* see Daley and Bergström (2012) and Daley et al. (2013). Predatory trilobites were documented by Bicknell et al. (2021).

16. For a summary see Vermeij (2016a). The unnamed cteacanthiform shark was discussed by Maisey et al. (2017).

17. For the radiodonts see Vinther et al. (2014) and Van Roy et al. (2015). The case for plankton-feeding in endocerids was made by Mironenko (2020) and endorsed by Stiefel (2021). *Titanichthys* was discussed by Coatham et al. (2020).

18. For a summary see Vermeij (2012a).

19. For the large dicynodont see Sulej and Niedźwiedcki (2019), *Patagosaurus* was described by Carballido et al. (2017). These authors consider higher mass estimates for titanosaurs like *Argentinosaurus*, with a published estimate of 99 tons, to be too high, because dinosaur bones are relatively light in construction and because of the many uncertainties in such calculations. For other masses of dinosaurs see Benson et al. (2014).

20. For the long gap between megaherbivores of the Late Cretaceous and Late Eocene see Onstein et al. (2022). See Larramendi (2016) for a careful and thorough review of giant elephants. Previous to his work, the largest post-Cretaceous mammal was thought to be the Late Oligocene Asian rhinoceros *Paraceratherium transouralicum*, standing 4.5 meters at the shoulder and weighing 10–20 tons. Larramendi prefers the lower estimate.

21. Benson et al. (2014), Tanaka et al. (2021).

22. Sereno et al. (2001), Aurellano et al. (2015).

23. For masses see Sorkin (2008) and Manzuetti et al. (2020).

24. For Paleozoic insects see Dorrington (2016) and Cannell (2018). For *Titanus* see the summary by Dvořáček et al. (2020).

25. For the pterosaur see Witton and Habib (2010). Bird sizes are from Chatterjee et al. (2007) and Vizcaíno and Fariña (1999).

26. For general accounts of cooperative hunting see Andersson (2005) and Bailey et al. (2013). The spotted hyena and African lion are thoroughly described by Kruuk (1972) and Schaller (1972) respectively. For cooperative hunting by ants see Dejean et al. (2005) and M. Schmidt and Dejean (2018). Data for the Aplomado falcon are from Hector (1986); for other cases in birds see Collopy (1983) and J. Jones (1999). For spiders see Yip et al. (2008) and Grinsted et al. (2020).

27. For the concept of the superorganism see Reeve and Hölldobler (2007), Strassmann and Queller (2007), and Hölldobler and Wilson (2009). Data on *Oecophylla* are from Pinkalsky et al. (2015).

28. For social insects see EO Wilson (2012).

29. For gregariousness in dinosaurs see Titus et al. (2021). Feeding by sabertooth cats is well described by Slater and Van Valkenburgh (2008) and Carbone et al. (2009).

30. For summary see Vermeij (2017b).

31. See Liston et al. (2013) for *Leedsichthys*. The history of pachycormids is documented by Friedman et al. (2010, 2013). For the history of plankton-feeding mysticetes see Tsai and Kohno (2016), Pyenson and Vermeij (2016), and Bisconti et al. (2020).

32. For the size of the living sperm whale see McClain et al. (2015). Estimates for *Otodus megalodon* come from Cooper et al. (2020) and Shimada (2021). These size estimates are somewhat lower than earlier approximations. Comparisons with the great white shark are given by Wroe et al. (2008). For the Mesozoic marine reptiles see Fröbisch et al. (2013) and Zverkov and Pervushov (2020). Recent estimates of maximum sizes of pliosaurs are again more conservative than some earlier claims. The Early Triassic ichthyosaur, *Cymbospondylus youngorum*, living 246 million years ago was even larger (estimated length 15 meters), but it was not an apex predator (Sander et al., 2021).

33. Summaries by Vermeij (2016a) and Pyenson and Vermeij (2016).

34. For habits of the killer whale see Pitman and Durban (2012). The hypothesis that killer whales were responsible for predation of very large whales is due to Springer et al. (2003).

35. Vermeij (2011b).

36. Vermeij (2012b, 2016a).

37. Burness et al. (2001).

Chapter 5: The Evolution of Motion

1. For the effects of vertical migration on redistribution of nutrients see Katija and Dabiri (2009) and Houghton et al. (2018). For the general case of locomotion and mixing see the excellent paper by Butterfield (2018). Thayer (1979, 1983) studied bioturbation and its stimulatory effects on productivity; see also the important work by Tarhan and her colleagues (2021) on the greater availability of phosphorus in bioturbated sediments.

2. For the general argument see Vermeij (1987). Defenses of passive sinking were documented by Kerfoot (1978) and Kerfoot et al. (1980). For the mosquito see Muijres et al. (2017).

3. For prokaryote locomotion see J. G. Mitchell (2002), Harshey et al. (2003), Polin et al. (2009), Stocker and Durham (2009), and Giblansky et al. (2010).

4. For an excellent summary of Ediacaran animals see Droser et al. (2017). Early locomotion was described in papers by A. G. Liu et al. (2010) and S. D. Evans et al. (2019).

5. Early burrowing and the subsequent spread of sediment bioturbation are documented by Droser et al. (2002), Dzik (2005), and Tarhan et al. (2015).

6. See Signor and Vermeij (1994) and especially C. D. Whalen and Briggs (2018). The unnamed cephalopod was discussed by Hildenbrand et al. (2021).

7. Klug and colleagues (2010) suggested that there was a marked increase in the diversity of swimming animals during the Devonian. They labeled this increase the Devonian Nekton Revolution. A more quantitative study by C. D. Whalen and Briggs (2018) indicated that the diversity of active swimmers increased gradually throughout the Paleozoic, from less than 1% of the fauna during the Cambrian to near 50% in the Late Paleozoic. Neither study considered swimming performance, which also seems to have increased during the era, although the pace and pattern of this increase remain unclear.

8. For bioturbation see Thayer (1983). Sand-burrowing, with special emphasis on streamlining of the shell, was discussed by Vermeij (2017b).

9. For calculated ichthyosaur speeds see Motani (2002). The evolution of fast fishlike swimming vertebrates is documented in papers by Donley et al. (2004), Wardle et al. (1989), Bernal et al. (2001), and Motani and Vermeij (2021). For fish speeds see Beamish (1978) and Iosilevskii and Weihs (2008).

10. For shell-bearing cephalopods and the locomotor history of cephalopods see Wells and O'Dor (1991), Chamberlain (1991), and Jacobs (1992). For the jumbo squid see Benoit-Bird and Gilly (2012).

11. For data on the silver ant and other fast terrestrial arthropods see Pfeffer et al. (2019). Avery and colleagues (1987) studied speeds of *Lacerta viridis*.

12. For *Eudibamus* see Berman et al. (2021). Limb postures in Paleozoic tetrapods were documented by Blob (2001).

13. See Sellers and Manning (2007) and Navarro-Lorbés et al. (2021) for speeds. Other recent estimates are considerably lower, especially for *Tyrannosaurus*, where speeds not exceeding two meters per second have been inferred from anatomy (Hutchinson and Garcia, 2002; van Bijlert et al., 2021). Bipeds are faster than their quadruped ancestors (Persons and Currie, 2017), but this does not mean that they were consistently fast; see discussion by Janis and Wilhelm (1993).

14. For speeds see Garland (1983), A. M. Wilson et al. (2013) and, for earlier Cenozoic mammals, Van Valkenburgh (1985).

15. For mammal speeds see Garland (1983). Fast extinct North American predators are discussed by Van Valkenburgh et al. (1990). Bramble and Lieberman (2004) document human adaptations to endurance running in humans.

16. For the absence of South American cursorial herbivorous mammals, see Croft and Lorente (2021). These authors suggest that this absence reflects phylogenetic constraints on the body plans of these extinct mammals. They point to phorusrhacid birds—South America's pursuit predators—as not having elicited rapid running adaptations in contemporaneous mammals. It is notable, however, that there were no long-distance pursuers among the native mammals. To me, "phylogenetic constraint" simply means that we do not understand why a given trait did not evolve.

17. For further discussion see BroJørgensen (2013).

18. See Trueblood and Seibold (2014) for squids, Rieppel et al. (2008) and Noè et al. (2017) for marine reptiles, Ruxton and Wilkinson (2011) for sauropods, Schaeffer and Rosen (1961) and Bellwood (2015) for jaw protrusion in fishes, Mehta and Wainwright (2007) for the moray eel, and Kohn (1956), Olivera (2002), and Schulz et al. (2019) for cone snails.

19. Comparisons among locomotor modes are from Hirt et al. (2017), who also point out that maximum speeds are achieved by medium-sized rather than by very large animals. For top speeds see McCracken et al. (2016). The endurance of the godwit during migration is documented by R. E. Gill et al. (2009). For the hummingbird see F. B. Gill (1985).

20. For the condor see H. J. Williams et al. (2020). Other soaring birds are documented by Sachs et al. (2013) and Weimerskirch et al. (2016).

21. For the earliest glider see E. Frey et al. (1997); for other gliders see Dudley and DeVries (1990).

22. For insects see Hasenfuss (2008) and Yanoviak et al. (2005, 2009); for birds and bats see Pei et al. (2020) and S. C. Anderson and Ruxton (2020); for pterosaurs, Baron (2021) and Ezcurra et al. (2020).

23. For flying squids see O'Dor et al. (2013) and especially Muramatsu et al. (2013). For flying-fish evolution see G.-H. Xu et al. (2015), Dasilao and Sasaki (1998), and Lewallen et al. (2011).

24. LaBarbera (1983), Conway Morris (2003). The ant case is documented by Grasso et al. (2020); for other terrestrial cases see the summary by Vermeij (2015a).

25. Sigwart et al. (2019).

26. For the long-term increase in activity see Bambach (2002).

Chapter 6: The Evolution of Violence

1. For multiple origins of mineralized skeletons see Zhuravlev and Wood (2008) and C. Liu et al. (2021). For predation and healed injuries see M. Y. Zhu et al. (2004), Z. F. Zhang et al. (2011), Peel (2015), and Bicknell et al. (2022).

2. For defenses of Cambrian trilobites see Ortega-Hernández et al. (2013), Dai et al. (2019), Pates and Bicknell (2019), Eriksson and Horn (2017), and Geyer et al. (2020). Enrollment was discussed by Vermeij (1987).

3. Vermeij (2020).

4. Conditions required for biomineralization are discussed by Wood et al. (2017). For oxygen being available only locally during the Proterozoic and even the Early Cambrian see D. B. Cole et al. (2020), Ding et al. (2019), and Tostevin et al. (2016). For an excellent review of biomineralization in animals see Gilbert et al. (2022).

5. For details see Vermeij (2015b); the idea was introduced in my 1975 paper. For the evolution of the gastropod operculum see Stanley (1982) and Checa and Jiménez-Jiménez (1998).

6. For the evolution of jaws see Brazeau and Friedman (2015) and Mironenko (2021). Prey adaptations were discussed by Cowen (1981) and Signor and Brett (1984).

7. See Vermeij (1975, 1977) for the escalation hypothesis, with special reference to the Late Mesozoic; and Vermeij (2022a) for ventrally weighted gastropods. Dietl (2003, 2004) has documented escalation between shell-entering busyconine gastropod whelks and tightly closing bivalves, also discussed by Vermeij (2015a, b). Additional cases of forcible prey entry are beautifully documented by Herbert et al. (2017) and Herbert (2018). Shell repair over time was documented by Vermeij et al. (1981). For the earliest shell-crushing crab see Ossó (2016).

8. For puffers see Vermeij and Zipser (2015); for loggerhead turtles see Marshall et al. (2012).

9. For *Dunkleosteus* see the functional analysis of P.S.L. Anderson and Westneat (2007). Estimates for Basilosaurus are from Snively et al. (2015); those for sharks are due to Wroe et al. (2008).

10. For the coconut crab in Okinawa see Oka et al. (2016). Estimates for crocodile bite forces are from Erickson (2012b) and Aurellano et al. (2015). Gignac and Erickson (2017) gave estimates for *Tyrannosaurus*; the role of the neck was suggested by Snively et al. (2014). Allosaurid jaws and skulls were described by Rayfield et al. (2001) and Frazzetta and Kardong (2002).

11. For the jaguar see Moral Sachetti et al. (2011); data for *Smilodon* are from Wroe et al. (2008). Bone cracking was discussed in fine papers by Ferretti (2007), Tseng and Wang (2010), Tseng et al. (2011), and Valenciano et al. (2016).

12. For early chewing see Rybczynski and Reisz (2001) and Angielczyk (2004). Hadrosaur dentition and chewing were documented by Erickson et al. (2012a). For the origin of chewing and the function of multicuspid teeth in mammals see Bhullar et al. (2019) and Virot et al. (2017).

13. For *Potamotrigon* see Kolmann et al. (2016); for *Siren* see Schwarz et al. (2021).

14. For sauropod dentition see Melstrom et al. (2021). Dinosaur diets are discussed by Weishampel and Jianu (2000), Hummel et al. (2008), and Fritz et al. (2011). Fritz and colleagues argued that oral preparation is not more efficient than breakdown further along in the digestive system. Efficiency, however, seems to be much less important than the quantity of food being ingested, which must be greater when it is mechanically masticated before being swallowed. For herbivorous birds see E. S. Morton (1978) and Dudley and Vermeij (1994). For the Messel birds and herbivory see Mayr (2017).

15. For a general review of predation on gastropods see Vermeij (2015c). Sea otter dentition and jaw mechanics are documented in Timm-Davis et al. (2017). For *Kolponomos* see Tseng et al. (2016). Shell-crushing was the likely trophic specialization of *Natusuchus* (Salas-Gismondi et al. (2015). Gans and De Vree (1986) investigated jaw mechanics in shell-crushing lizards.

16. Vermeij (2015c).

17. For *Coris* see Vermeij and Zipser (2015); for further discussion of pharyngeal bones in fishes see Yamaoka (1978) and McGee et al. (2016).

18. For the oldest known seed-eating rodent see Collinson and Hooker (2000); for other seed-eaters, see Morse (1975) for birds, and Thiery and Sha (2020) for birds and primates. For the evolution of nut-eaters and nut-producing trees see Vander Wall (2001), Tomback and Linhart (1990), B.-F. Zhou et al. (2022) (Fagaceae), and H. Zhou et al. (2021) (Juglandaceae). Tiffney (1984) and Friis et al. (2015) document the increase in seed size from the Cretaceous to the Cenozoic.

19. For feeding habits of primates see Kay (1981) and Thiery and Sha (2020); for birds see Schrorger (1941) and Willson (1972); for peccaries see Kiltie (1982). Cognitive abilities of hoarders were studied by Raby et al. (2007).

20. For the jaguar see Emmons (1989) and Miranda et al. (2016).

21. For the hypothesis that defense played a major role in the evolution of eusociality see Nowak et al. (2010) and E. O. Wilson (2012).

22. For a general review of venomous animals and the genetics of venoms see Sunagar and Moran (2015). Fish venoms and their functions are nicely summarized by W. L. Smith et al. (2016). The ancestral condition and function of stings in Hymenoptera were inferred phylogenetically by B. R. Johnson et al. (2013) and Peters et al. (2017).

23. For caecilians see Mailho-Fontana et al. (2020). Conoidean envenomation and history are discussed by Shimek and Kohn (1981) and Kantor and Puillandre (2012). For Colubrariidae see Oliverio and Modica (2010). The hypothesis that all snakes (venomous as well as nonvenomous) and venomous lizards belong to a single clade and evolved envenomation only once was proposed by Fry et al. (2006).

24. For nettles, deer, and insects see Iwamoto et al. (2014). The Early Eocene nettle was described by DeVore et al. (2020).

25. For a fine review of electric fishes see Crampton (2019). For the electric eel see Bastos et al. (2021).

26. See the excellent paper by Roach et al. (2013) for human adaptations to throwing. William McNeill's (1982) book on the evolution of human weapons is a classic.

27. For ants see E. O. Wilson (2012). Harrowing accounts of group warfare in the banded mongoose are given by Green et al. (2022).

28. The weapons-for-show argument was most eloquently elaborated by Emlen (2008). Valerius Geist's intensive fieldwork on moose sheep and other ungulates with head gear demonstrated high frequencies of injury despite the mating-related ritualistic combat in which these animals also engage; see for example his 1978 paper.

29. Pinker (2011).

30. See for example Lev-Yadun (2001, 2014, 2016); and a discussion by Vermeij (2016b).

31. Land-to-sea transitions in tetrapods were documented and discussed by Vermeij and Motani (2018).

Chapter 7: The Power of Economies

1. See Canfield et al. (2006), Ward et al. (2019), and discussion by Vermeij (2019b).

2. See Butterfield (2015), Crowe et al. (2013), and especially Fournier et al. (2021) for the Archean origin of photosynthesis and cyanobacteria. Estimates of Late Archean productivity are from Hao et al. (2020) and Canfield (2021). For the continued role of prokaryotes as primary producers see Brocks et al. (2017). Whether oxygen-releasing photosynthesis—and cyanobacteria—arose in freshwater rather than the sea remains contentious; for a recent analysis, which favors a freshwater origin, see Raven and Sánchez-Baracaldo (2021). For the timing of the Great Oxidation Event see Gumsley et al. (2017).

3. For a thorough account see van der Giezen and Lenton (2012).

4. For the rise of green algae see Brocks et al. (2017).

5. For the early spore record see Strother and Foster (2021). Adiatma et al. (2019) suggested a link between plants and a rise in oxygen during the Middle Ordovician. For an excellent summary of early vascular plants see C. J. Harrison and Morris (2018). For the evolution of Paleozoic soils occupied by bryophytes see R. L. Mitchell et al. (2021).

6. For the evolution of leaves see C. J. Harrison and Morris (2018). Beerling et al. (2001) proposed that leaves could not evolve before carbon dioxide levels began to fall in the Late Devonian, a position also mentioned by Harrison and Morris.

7. For the evolution of roots see Hetherington and Dolan (2018). Ligrone et al. (2012) discussed mosses and their anatomy.

8. Increasing oxygen levels to modern levels are indicated by a higher frequency of charcoal in Late Devonian sediments (Glasspool et al., 2015) and by isotopic data (Wallace et al., 2017). The argument that low leaf-vein density in Paleozoic plants implies low productivity is due to Boyce et al. (2009) and was elaborated by Boyce and Leslie (2012) and Boyce and Zwieniecki (2019). J. P. Wilson et al. (2017, 2020) have contested this view, arguing that high conductance

of stems in Paleozoic plants indicates high productivity, and that modern plants with low productivity are not good models for interpreting Paleozoic lycophytes and pteridosperms. The fact that the foliage of Carboniferous plants tends to be thick and sometimes very large indicates that these leaves stayed on the plants for long periods, perhaps several years. These resistant leaves do not decay quickly and are therefore susceptible to burial as peat and coal without being fully recycled (J. Robinson, 1990). To me, the available evidence favors the hypothesis of low productivity. For additional anatomical evidence of low productivity in Carboniferous tree-like lycophytes see Carriquí et al. (2019) and D'Antonio and Boyce (2020). For the early history of terrestrial herbivory see Labandeira (2013), Labandeira et al. (2014), and Q. Xu and Labandeira (2018). Boyce and Knoll (2002) document the evolution of net-veined leaves. The connection between low productivity and the scarcity of herbivores is my own.

9. See Boyce et al. (2009), Boyce and Leslie (2012), Boyce and Lee (2010, 2011), and Brodribb et al. (2013). For the role of herbivory see Vermeij (2017a).

10. For early nitrogen fixation see Garvin et al. (2009) and Canfield et al. (2010). The oldest fossil fabaceous pod was documented by Centeno-González et al. (2021). Koenen et al. (2021) document the phylogeny of Fabaceae. Epihov et al. (2017, 2021) investigated the role of nitrogen-fixing Fabaceae in various vegetation types.

11. Owen (1980) was the first to suggest that grasses owe their success to herbivorous mammals. For the phylogeny of grasses see Bouchenak-Khelladi et al. (2014). The sheath-like basal part of the grass leaf is an important innovation derived from the petiole in other angiosperms (A. E. Richardson et al., 2021). It is unclear when this innovation appeared, but basal growth of leaves is a widespread characteristic of monocots, a major angiosperm clade whose divergence from other flowering plants is estimated to have occurred 131 million years ago during the Early Cretaceous (Hertweck et al., 2015); see also Chase (2004) for interrelationships among monocots. Most early monocots still had petiolate leaves, as is the case in living members of such families as Araceae and Dioscoreaceae.

12. For these events see respectively Brocks et al. (2017), Del Cortona et al. (2020), and LoDuca et al. (2017).

13. See Falkowski et al. (2004), Sinninghe Damsté et al. (2004), Knoll and Follows (2016), Eichenseer et al. (2019). The interpretation that herbivores might have triggered these transitions is mine, and conflicts with the contention by Knoll and Follows that the Jurassic changes largely reflect a competitive advantage to larger phytoplankters. Sharoni and Halevy (2022) suggested that the modeled increase in phosphorus content relative to carbon and nitrogen, and therefore the success of the red-algal-based phytoplankters, is attributable to the breakup of the supercontinent Pangaea after the Middle Triassic. This would have increased terrestrial weathering of phosphorus and increased availability of phosphorus in the ocean. Why the much earlier breakup of the supercontinent Rodinia about the Proterozoic-Cambrian boundary would not also have increased the phosphorus content of phytoplankton is unclear. Sharoni and Halevy hint at the possibility that bioturbation and increasing activity on land might be partly responsible for the greater availability of phosphorus for Mesozoic and later phytoplankters. This suggestion accords well with the likelihood that the higher post-Triassic productivity of the phytoplankton is at the very least magnified by biological feedbacks.

14. For turf algae see Welsh and Bellwood (2014). For large leafy temperate algae and seagrasses, see the summary in Vermeij (2017a). Nitrogen fixation in the seagrass *Posidonia* was documented by Mohr et al. (2021). For mangrove and salt-marsh plants see Vermeij (2022a).

15. For the nature of the feedbacks see Bullen et al. (2021); for the history of North Pacific kelps and mammals see Vermeij (2018b).

16. For general background see Vermeij (1987, 2004b); Bambach (1993); Bambach et al. (2002); and Allmon et al. (2014).

17. S. J. Gould and Calloway (1980) held that the general replacement of brachiopods by bivalves was the consequence of the greater susceptibility of brachiopods to extinction during the end-Permian crisis and considered the replacement to reflect the much faster rebound of bivalves. I showed, however, that the decline of brachiopods only began during the Early Cretaceous and argued that bivalves tended to exclude brachiopods from most shallow-water environments by virtue of their higher metabolic rates (Vermeij, 1987). Steele-Petrovic (1979) had already amassed evidence for the lower pace of life of brachiopods. The idea that some fossil Paleozoic brachiopods were photosymbiotic was first proposed by Cowen (1970, 1983) and further elaborated by *Gigantoproductus* by Angiolini et al. (2019). Stanley (1968) documented the increasing strength of ciliary currents in bivalves as well as the proliferation of siphonate lineages.

18. For crinoids see the very clever work by Saulsbury (2020).

19. Chan et al. (2021).

20. See Vermeij (2002) for shell elaborations, and Finnegan et al. (2011) for increases in metabolic rates in gastropods.

21. For insects see Heinrich (1993); for vertebrates see Ruben (1995), Lovegrove (2017), Knaus et al. (2021), and Chiarenza et al. (2022). Grigg and colleagues (2022) suggested that full (or whole-body) endothermy was an ancestral condition in early terrestrial amniotes (the group including reptiles, birds, and mammals), and that it was subsequently lost in sauropod dinosaurs as well as in the ancestors of lizards, snakes, and crocodiles. In their view, endothermy arose just once in amniotes rather than separately in the ancestors of mammals, pterosaurs, and theropod dinosaurs including birds. Early amniotes likely had higher metabolic rates than their amphibian-like progenitors, as indicated by enhanced blood flow in long bones and in the erect posture and gait of some lineages, but these features are not consistently diagnostic of endothermy, and some structures associated with endothermy in birds and mammals, such as respiratory turbinates that reduce water loss during air-breathing, are absent in Paleozoic lineages. The early presence of suitable muscle physiology and some other anatomical traits in early amniotes might well have predisposed descendant lineages to evolve mammal-like or bird-like endothermy, but by itself does not imply a high body temperature generated by heat production. Reversion to ectothermy is unknown in birds and is very rare in mammals. Loss of endothermy has evidently occurred in the Namibian desert golden moles of the genus *Eremitalpa*, which live under arid conditions where productivity is very low. It strains my credulity to think that the ancestors of sauropods, snakes, lizards, and crocodiles would have lived in such unproductive environments where endothermy would be a liability. Some of the latter animals could have become gigantotherms, but I adhere to the more conventional view (Lovegrove, 2017) that they, and most Permian reptiles, were effectively ectotherms. Evidence for endothermy from

characters of the vertebrate inner ear comes from a compelling study by Araújo et al. (2022). In the view of these authors, endothermy in the mammal lineage arose just once, and much later than proposed by Grigg et al. (2022).

22. See discussion in Vermeij (2015a).

23. For claims of endothermy in Mesozoic marine reptiles see A. Bernard et al. (2010), Houssaye et al. (2013), Fleischle et al. (2018), and Lindgren et al. (2018). The benefits of marine endothermy at high latitudes were outlined by Cairns et al. (2008) and Grady et al. (2019). Lovvorn (2010) discussed the smallest marine endotherms (auklets weighing 44 grams).

24. For the early appearance of wood-digesting fungi see Nelsen et al. (2016), Hibbett et al. (2016), and Ayuso-Fernández et al. (2018). For arthropods, including termites, and other important animals as decomposers of wood and decaying plant litter in soil, see Frouz (2018) and Ulyshen (2016). For further discussion of ecosystem efficiency see Vermeij (2019a).

25. Robinson (1990).

26. See Bowman et al. (2009), Belcher et al. (2021), and Bond and Midgley (2012).

27. See Vermeij (2019a, b) for the general case for increasing power over time. Kidwell and Brenchley (1996) documented increasing thicknesses of shellbeds over the Phanerozoic. For the silica and calcium cycles see Kidder and Erwin (2001) and Ridgwell and Zeebe (2005) respectively.

28. For decreasing background extinction in relation to oxygen see Vermeij (2019a) and Stockey et al. (2021). Bachan et al. (2017) and Payne et al. (2020) document increasing regulation of the carbon cycle over time.

29. Polis et al. (1997).

30. For summaries see Vermeij (1991, 2005).

31. Vermeij and Motani (2018).

32. For the common good see Hardin (1968), Vermeij and Leigh (2011), Vermeij (2018a).

33. Vermeij (2019a). Lotka (1922) suggested that efficiency is greatest where energy is most limited. In my interpretation this statement applies to individual organisms but not to ecosystems. Por (1994) suggested that processes in ecosystems have become increasingly efficient, but he did not elaborate on this idea, and it is not quite clear what he meant by efficiency. Many biologists have confused efficiency with effectiveness; I equate the latter concept with performance.

34. James Estes and colleagues beautifully documented the role of sea otters and sea urchins in kelp communities in the Aleutian Islands of Alaska; see for example Estes and Palmisano (1974) and Estes et al. (2016). For the feedbacks involving seacows and kelps see Bullen et al. (2021).

35. Savoca et al. (2021), Smetacek (2021).

36. Karp et al. (2021).

37. Vermeij (2019a, b).

38. There is a large body of literature indicating that higher diversity promotes primary productivity; see for example Tilman et al. (2012) and Walde et al. (2021). Experiments with plant species diversities ranging from 1 to 20 species sown in plots in temperate grasslands in Minnesota and Switzerland show that plots with more species are more productive. These studies suffer from three problems. First, the number of species, even in the most diverse plots, is

small, leaving open the question whether much higher numbers of species would yield still more production. Second, the communities monitored consisted only of plants; feedbacks on productivity established by grazing were not considered, although Tilman and colleagues argue that herbivory is a less important contributor to productivity than is the number of plant species. Third, the conclusion that increased diversity "explains" higher productivity is an example of assigning cause to a descriptive relationship when correlation between two phenomena would have been a more cautious interpretation. To my mind, small-scale experiments in simplified artificial communities can yield misleading results. Despite my reservations, these studies, especially those by David Tilman and his colleagues, have been enormously influential in ecology.

39. A. Smith (1776).

40. For forests see Cristoffer and Peres (2003); for Caribbean reefs see K. G. Johnson et al. (2008).

41. For geographical patterns in diversity see Vermeij (2012a, 2018b).

42. The power-law distribution of inequality is explored by Fix (2018, 2019) and Scheffer et al. (2017). The rise of inequality from hunter-gatherers to sedentary societies was documented by Borgerhoff Mulder et al. (2009). For additional discussion of inequality in human societies see the excellent books by Piketty (2014, 2020) and Reich (2015), as well as the fine historical paper by Alfani (2021).

43. Vermeij (2004a), Roopnarine (2006). For a review of mass extinctions and their causes see D.P.G. Bond and Grasby (2017). Many studies assign a primary causal role to anoxia and acidification; for a recent paper see Fox et al. (2022).

44. For post-Cretaceous restoration see D'Hondt et al. (1998), Lowery et al. (2018), Brosse et al. (2019), Henehan et al. (2019), and Lyson et al. (2019). The slower recovery of powerful organisms and feedbacks can be inferred from papers by Henehan et al. (2019), Lyson et al. (2019), Rodríguez-Tovar et al. (2020), and C.P.A. Smith et al. (2021). For the post-Permian see Erwin (1996) and especially Roopnarine et al. (2007) and Roopnarine and Angielczyk (2015). Delayed recovery has generally been attributed to multiple post-crisis disruptions occurring in rapid succession (S. M. Stanley, 1990; Hallam, 1991), but I wonder whether a more robust ecosystem with better-developed feedbacks would have been as susceptible to such disruptions.

45. Vermeij (2008b).

46. For increasing brain size see Bertrand et al. (2022).

Chapter 8: The Human Singularity

1. For a summary of anthropoid and hominid evolutionary history see Brunet and Jaeger (2017). Important papers on human brain evolution are by Neubauer et al. (2018), Raghanti et al. (2018), and Ponce de León et al. (2021). For other anatomical adaptations see Bramble and Lieberman (2004), Lieberman (2007), and Roach et al. (2013). For the early emergence of cooking food see Wrangham (2017). See also the chapter by Vermeij (2023).

2. See Blasi et al. (2019) for this example. The relationship between mechanics and anatomy has long been recognized. D'Arcy W. Thompson (1942), for example, was so convinced by this link that he doubted the role of natural selection in the adaptive process. I suggested in 2002

that forces imposed by the environment and by the animal itself influence the evolution of shell shape in molluscs. For the general case of a facultative link between form and force or behavior becoming genetically obligate, see the fine papers by Palmer (2004) and G. E. Robinson and Barron (2017). For Dutch speakers I highly recommend the iconoclastic book by Oudman and Piersma (2018).

3. For a discussion of ultrasociality and the conditions necessary for its evolution see D. S. Wilson and Gowdy (2015). There is a well-entrenched tradition in evolutionary biology denying the importance of groups in evolution (see for example G. C. Williams, 1966). This view stems from the perception that evolutionary adaptation can come only from changes in gene frequencies. Under that assumption, selection at group level is indeed rare (Leigh, 1983). Given that cultural traits can be inherited and that groups can under some circumstances act as individuals, the conditions for group properties to be selected and to influence evolution become less stringent.

4. For further discussion see Buss (1987), West et al. (2015), and Leigh (2010).

5. For the role of common enemies see Levin (1999). Warfare and cooperation were explored by Bowles (2009). For nonhuman warfare see Crofoot and Wrangham (2010).

6. Vermeij (2023).

7. Faurby et al. (2020).

8. Vermeij (2023).

9. Diamond (1997). For horse domestication and the invention of the wheel see Anthony (2007). The origin of modern science is treated by Strevens (2020), whose work is also the subject of the next paragraph.

10. Many economists and political scientists have written about trust. For a fine recent account see Seabright (2004). Ferguson (2008) chronicled the history and uses of money, including the importance of financial institutions in spreading risk and promoting trust. Reich (2015) and Piketty (2014, 2020) have made a strong case for the importance of diminishing social inequality as a means to increase trust in the economic sphere.

11. Syvitski et al. (2020); also see Brown et al. (2011). For the technosphere see Zalasiewicz et al. (2017).

12. See Vermeij and Leigh (2011).

13. Ibid.

14. For early effects by humans on Earth's landscape see ArchaeoGlobe Project (2019).

15. For good histories of the transition out of the Malthusian trap see G. Clark (2007), Harari (2015), and Landes (1998). See also Pinker (2011) for a particularly optimistic view.

16. For critiques of the free market see Stiglitz (2002), Reich (2015), and D. S. Wilson and Gowdy (2015). For the rarity of monopolies in nature see Vermeij and Leigh (2011). In his fine book on liberalism, Fukuyama (2022) makes it plain that both capitalism and the identity politics of race, gender, and ethnicity tilt the distribution of opportunity in favor of elites. Somewhat tellingly, perhaps, he does not consider disability in his list of disadvantaged classes. Although Fukuyama does not express his views in terms of power, it is power that distinguishes elites from the rest of human society.

17. Many studies have grappled with the rise of the Nazi party. Some accept religious affiliation as a key factor (Spenkuch and Tillmann, 2018); others point to the corrosive effect of

government austerity during the tenure of the Weimar Republic (Galofré-Vilà et al., 2021), and still others the role of friends in civic social associations as the key to the spread of populist ideas (Satyanath et al., 2017). On the other hand, Boeri and collegues (2021) reinforce the more conventional view that membership in social organizations offers some protection against acceptance of Communist or Fascist ideologies that pit the virtuous people against the corrupt elite. I suspect that dangerous ideas spread both because the originators are especially powerful in their advocacy and because society has either lost or never had the resistance or the workable alternatives to the authoritarian ideologies. Either way, there is a power differential. It's complicated.

18. Dennett (2006) has emphasized the power of early childhood exposure in perpetuating religious teachings.

19. A. Smith (1776), D. S. Wilson and Gowdy (2015).

20. Hardin (1968).

21. Vermeij (2018a).

22. For the benefits of growth and a generally optimistic view of its continuation see Lewis (1955), Romer (1986), Kremer (1993), and Pinker (2018). Interestingly, these authors rarely mention the possibility of misallocating the wealth that results from economic growth.

23. See Czech (2019), who does not discuss the failure to account for the accumulated technosphere.

24. For the shallow-to-deep, onshore-offshore trend see Jablonski et al. (1983), Jablonski and Bottjer (1990).

25. Simon (1977), Reich (2015), Piketty (2020).

26. See Fukuyama (2022). Like most philosophers, Fukuyama does not stray into the realm of nonhuman life, but his analysis of liberalism and its critics is clear and sound.

27. See for example T. Jackson (2017), Pinker (2018), and especially Murtaza (2011).

28. For the rise of the welfare state see Lindert (2004). I agree with Hammond (2019) that philanthropy should not replace government-sponsored welfare: In the Middle Ages, when all welfare in Europe was in the hands of the Catholic Church, alms were woefully inadequate and important good works such as educating women and the disabled were not attempted.

Chapter 9: Power, Pattern, and the Emergence of Beauty

1. See Gould (1989, 2002). In his 1985 paper, Gould explicitly accorded adaptation and natural selection a subordinate role in evolution to chance events. Gould's view represents a nearly complete reversal of his early, strongly adaptationist writings. Padian (2017) points out that the importance of the end-Cretaceous extinction was likely exaggerated by Gould and others. Not only were mammals already quite diverse during the Cretaceous, but dinosaur diversity may have been declining well before the crisis.

2. Most evolutionary biologists have abandoned the use of the value-laden English terms in evolution, and have come to realize that many seemingly "primitive" or "simple" traits are in fact secondarily derived from states that would once have been considered "advanced." As noted in chapter 1, lineages that are on the losing side of competitive interactions initially lose power and undergo a reduction in expensive structures (eye loss in cave fishes, for example), but once

they adapt to their circumstances of diminished resources, they are subject to selection for as much power as conditions allow. McShea (2001) has pointed out that secondary loss of structures might be more common than gains. I interpret this fact to mean that there are many more ways of making a living as a lineage with little power than as one in which performance levels are high.

3. See Berlin (1954, 1979), Popper (1964), Monod (1971), and critical discussion by Vermeij (2011b).

4. Other proponents of a predominantly contingent interpretation of history include Monod (1971), Beatty (2006), Powell and Mariscal (2015), Losos (2017), and Blount et al. (2018). None of these thinkers, including Gould, considered that they were dealing with temporary states. In previous papers (Vermeij, 2006, 2019b) I also failed to recognize that the particulars of history reside in structures strictly bounded in time and space. Monod is explicit in considering contingency to apply to objects.

5. Kauffman (1993, 2000) is one of many scientists who have emphasized the importance of self-organization. For the role of this process in the origin of life see Lane (2009, 2015). The inherent strength of chemical bonds affects the three-dimensional structures of proteins and nucleic acids, implying that some configurations and functions of polymers are influenced by energetically favorable relationships among the component nucleotide bases and amino acids.

6. The idea of a positive feedback between selection and environmental favorability is due to Lenton (1998). The rephrasing and elaboration of this idea are mine.

7. Leigh and Ziegler (2019).

8. Conway Morris (2003) argued that historical directionality is demonstrated by repeated and separate origins of particular adaptive traits such as endothermy. He pointed to the very high frequency of convergence in separate lineages to similar adaptive states. In my studies (Vermeij, 2006, 2019b) I found that directionality emerges from selective processes that favor some innovations because those states benefit organisms under a wide range of conditions. In the 2019b paper, I emphasized the importance of performance levels and power in the emergence of historical directionality. For biomineralization see chapters 4 and 6. A remarkable terrestrial origin of biomineralization was recently described in attine leaf cutter ants (Li et al., 2020).

9. Vermeij (2006).

10. See the excellent books by S. B. Carroll (2005) and Wagner (2014).

11. For conditions on oceanic islands see Leigh et al. (2009).

12. For excellent books on human invention and technology see Mokyr (1990) and Arthur (2009). For the social receptivity of invention and science see Snooks (1998).

13. See Knoll et al. (1996, 2007) and Payne et al. (2016) for the role of tolerance to carbon dioxide in survival during the end-Permian crisis. For a discussion of unpredictability and adaptation see Vermeij (2008b).

14. Vermeij (1995).

15. See Prum (2017). The connection between beauty and power is my own.

BIBLIOGRAPHY

Abraham AJ, Webster AB, Jordaan J et al. 2022. Hyenas play unique ecosystem role by recycling key nutrients in bones. Afr. J. Ecol. 60: 81–86.

Adiatma YD, Saltzman MR, Young SA et al. 2019. Did early land plants produce a stepwise change in atmospheric oxygen during the Late Ordovician (Sandbian ~458 Ma)? Palaeogeogr., Palaeoclimatol., Palaeoecol. 534: 109341.

Albers S-V and Jarrell KF. 2015. The archaellum: how Archaea swim. Front. Microbiol. 6: 23.

Alexander RM. 1977. Swimming. In Alexander RM (ed.), Mechanics and energetics of animal locomotion. Chapman and Hall, London: 222–248.

Alfani G. 2021. Economic inequality in preindustrial times, Europe and beyond. J. Econ. Lit. 59: 3–44.

Allmon WD and Martin RD. 2014. Seafood through time revisited: The Phanerozoic increase in marine trophic resources and its macroevolutionary consequences. Paleobiology 40: 256–287.

Anderson PSL and Westneat MW. 2007. Feeding mechanics and bite force modelling of the skull of *Dunkleosteus terrelli*, an ancient apex predator. Biol. Lett. 3: 76–79.

Anderson SC and Ruxton GD. 2020. The evolution of flight in bats: A novel hypothesis. Mammal Rev. 50: 426–439.

Andersson K. 2005. Were there pack-hunting canids in the Tertiary, and how can we know? Paleobiology 31: 56–72.

Angielczyk KD. 2004. Phylogenetic evidence for and implications of dual origin of propaliny in anomodont therapsids (Synapsida). Paleobiology 30: 268–296.

Angiolini L, Crippa G, Azmy K et al. 2019. The giants of the phylum Brachiopoda: A matter of diet? Palaeontology 62: 889–917.

Antcliffe JB, Callow RHT, and Brasler PD. 2014. Giving the early fossil record of sponges a squeeze. Biol. Revs. 89: 972–1004.

Anthony DW. 2007. The horse, the wheel, and language: How Bronze-Age riders from the Eurasian steppes shaped the modern world. Princeton University Press, Princeton.

Araújo R, David R, Benoit J et al. 2022. Inner ear biomechanics reveals a Late Triassic origin of mammalian endothermy. Nature 607: 726–731.

ArchaeoGlobe Project. 2019. Archaeological assessment reveals Earth's early transformation through land use. Science 365: 898–902.

Arthur WB. 2009. The nature of technology: What it is and how it evolves. Free Press, New York.

Ashton BV, Freestone AL, Duffy JE et al. 2022. Predator control of marine communities increases with temperature across 115 degrees of latitude. Science 376: 1215–1219.

Aurellano T, Ghilardi M, Guilherme E et al. 2015. Morphometry, bite force, and paleobiology of the Late Miocene caiman *Purussaurus brasiliensis*. PLoS One 10: eon 7944.

Avery RA, Mueller CF, Jones SM et al. 1987. Speeds and movement patterns of European lacertid lizards: A comparative study. J. Herpetol. 21: 324–329.

Ayuso-Fernández I, Ruiz-Dueñas FJ, and Martínez T. 2018. Evolutionary convergence in lignin-degrading enzymes. Proc. Nat. Acad. Sci. U.S.A. 115: 6428–6433.

Bachan A, Lau KV, Saltzman MR et al. 2017. A model for the decrease in amplitude of carbon isotope excursions across the Phanerozoic. Amer. J. Sci. 317: 641–646.

Bailey I, Myatt JP, and Wilson AM. 2013. Group hunting within the Carnivora: Physiological, cognitive and environmental influences on strategy and cooperation. Behav. Ecol. Sociobiol. 67: 1–17.

Bambach RK. 1993. Seafood through time: Changes in biomass, energetics, and productivity in the marine ecosystem. Paleobiology 19: 372–397.

Bambach RK. 2002. Supporting predators; changes in the global ecosystem inferred from changes in predator diversity. In Kowalewski M and Kelley PH (eds.), The fossil record of predation. Paleont. Soc. Pap. 8: 319–351.

Bambach RK, Knoll AH, and Sepkoski JJ Jr. 2002. Anatomical and ecological constraints on Phanerozoic animal diversity in the marine realm. Proc. Nat. Acad. Sci. U.S.A. 99: 6854–6859.

Baron MG. 2021. The origin of pterosaurs. Earth-Sci. Revs. 121: 103777.

Bar-On MM, Phillips R, and Milo R. 2018. The biomass distribution on Earth. Proc. Nat. Acad. Sci. U.S.A. 115: 65-6-6511.

Bart MC, Hudspith M, Rapp HT et al. 2021. A deep-sea sponge loop? Sponges transfer dissolved and particulate organic carbon and nitrogen to associated faunas. Front. Mar. Sci. 8: 604879.

Bartusevičius H, van Leeuwen F, and Petersen MB. 2020. Dominance-driven autocratic political orientations predict political violence in Western, educated, industrialized, rich, and democratic (WEIRD) samples. Psychol. Sci. 31: 1511–1530.

Bastos DA, Ruanon J, Py-Daniel LR, and Santana CD de. 2021. Social predation in electric eels. Ecol. Evol. 11: 1088–1092.

Beamish FWH. 1978. Swimming capacity. In Hoar WS and Ryall DJ (eds.), Fish physiology, Vol. 7 (Locomotion). Academic Press, New York: 101–187.

Beatty J. 2006. Replaying life's tape. J. Philos. 103: 336–362.

Beerling DJ, Osborne CP, and Chaloner WG. 2001. Evolution of leaf-form in land plants linked to atmospheric CO decline in the Late Paleozoic era. Nature 410: 352–354.

Belcher CM, Mills BJW, Vitali R et al. 2021. The rise of angiosperms strengthened fire feedbacks and improved the regulation of atmospheric oxygen. Nature Comm. 12: 503.

Bellwood DR, Goatley CHR, Bellwood O et al. 2015. The rise of jaw protrusion in spiny-rayed fishes closes the gap on elusive prey. Curr. Biol. 25: 2696–2700.

Bengtson S and Zhao Y. 1992. Predatorial borings in Late Precambrian mineralized exoskeletons. Science 257: 367–369.

Benoit-Bird KJ and Gilly WF. 2012. Coordinated nocturnal behavior of foraging jumbo squid *Dosidicus gigas*. Mar. Ecol. Prog. Ser. 455: 211–228.

Benson RBJ, Campione NE, Carrano MT et al. 2014. Rates of dinosaur body mass evolution indicate 170 million years of sustained ecological innovation on the avian stem lineage. PLoS Biol. 12: e1001896.

Benton MJ. 1983. Dinosaur success in the Triassic: A noncompetitive ecological model. Quat. Rev. Biol. 58: 29–51.

Berlin I. 1954. Historical inevitability. Oxford University Press, London.

Berlin I. 1979. The concept of scientific history. In Concepts and categories: Philosophical essays. Viking, New York: 103–142.

Berman DS, Sumida SS, Henrici AC et al. 2021. The Early Permian bolosaurid *Eudibamus cursoris*: Earliest reptile to combine parasagittal stride and digitigrade posture during quadrupedal and bipedal locomotion. Front. Ecol. Evol. 9: 674173.

Bernal D, Dickson KA, Shadwick RE, and Graham JB. 2001. Analysis of the evolutionary convergence for high performance swimming in lamnid sharks and tunas. Comp. Biochem, Physiol. A 129: 695–726.

Bernard A, Lécuyer C, Vincent P et al. 2010. Regulation of body temperature by some Mesozoic marine reptiles. Science 328: 1379–1382.

Bertness PD. 1981. Conflicting advantages in resource utilization: The hermit crab housing dilemma. Amer. Nat. 118: 432–437.

Bertness PD. 1985. Fiddler crab regulation of *Spartina alterniflora* production on a New England salt marsh. Ecology 66: 1042–1055.

Bertrand OC, Shelley SL, Williamson TE et al. 2022. Brawn before brains in placental mammals after the end-Cretaceous extinction. Science 376: 80–85.

Bhullar B-A S, Manafzadeh AR, Miyamae JA et al. 2019. Rolling of the jaw is essential for mammalian chewing and tribosphenic molar function. Nature 566: 528–532.

Bianucci G, Marx FG, Collareta A et al. 2019. Rise of the titans: Baleen whales became giants earlier than thought. Biol. Lett. 15: 2019.0175.

Bicknell RDC, Holmes JD, Edgecombe GD et al. 2021. Biomechanical analyses of Cambrian euarthropod limbs reveal their effectiveness in mastication and durophagy. Proc. Roy. Soc. B 288: 2020.2075.

Bicknell RDC, Holmes JD, Pates S et al. 2022. Cambrian carnage: Trilobite predator-prey interactions in the Emu Bay Shale of South Australia. Palaeogeog., Palaeoclimatol., Palaeoecol. 591: 110877.

Bicknell RDC and Paterson JR. 2018. Reappraising the early evidence of durophagy and drilling predation in the fossil record: Implications for escalation and the Cambrian explosion. Biol. Revs. 93: 754–784.

Bijlert PA van, Soest AJ van, and Schulp AS. 2021. Natural frequency method: Estimating the preferred walking speed of Tyrannosaurus rex based on tail natural frequency. Roy. Soc. Open Sci. 8: 201441.

Birkeland CE. 1989. Geographic comparisons of coral-reef community processes. Proc. Sixth Intern. Coral Reef Symp. 1: 211–220.

Birkeland CE and Dayton PK. 2005. The importance in fishery management of leaving the big ones. Trends Ecol. Evol. 20: 356–358.

Bisconti M, Munsterman DK, Fraaije RHB et al. 2020. A new species of rorqual whale (Cetacea, Mysticeti, Balaenopteridae) from the Late Miocene of the southern North Sea Basin and the role of the North Atlantic in the paleobiogeography of Archaebalaenoptera. PeerJ 8: e8315.

Blasi DE, Mroan S, Moisik SR et al. 2019. Human sound systems are shaped by post-Neolithic changes in bite configuration. Science 363: eaav3218.

Blob RW. 2001. Evolution of hindlimb posture in nonmammalian therapsids: Biomechanical tests of paleontological hypotheses. Paleobiology 27: 14–38.

Block BA, Finnerty JR, Stewart AFR, and Kidd J. 1993. Evolution of endothermy in fish: Mapping physiological traits on a molecular phylogeny. Science 260: 210–214.

Blount, ZB, Lenski RE, and Losos JB. 2018. Contingency and determinism in evolution: Replaying life's tape. Science 362: eaam5979.

Bobrovskiy I, Hope JM, Nettersheim BJ et al. 2021. Algal origin of sponge sterane biomarkers negates the oldest evidence for animals in the rock record. Nature Ecol. Evol. 5: 165–168.

Boeri T, Mishra P, Papageorgiou C, and Spilimbergo A. 2021. Populism and civil society. Economica 88: 863–895.

Bond DPG and Grasby SE. 2017. On the causes of mass extinctions. Palaeogeogr., Palaeoclimatol., Palaeoecol. 478: 3–29.

Bond WJ and Midgley JJ. 2012. Fire and the angiosperm revolutions. Intern. J. Plant Sci. 173: 569–583.

Borgerhoff Mulder M, Bowles S, Hertz T et al. 2009. Intergenerational wealth transmission and the dynamics of inequality in small-scale societies. Science 326: 682–688.

Bouchenak-Khelladi Y, Muasya AM, and Linder HP. 2014. A revised evolutionary history of Poales: Origins and diversification. Bot. J. Linn. Soc. 175: 4–16.

Bowles S. 2009. Did warfare among ancestral hunter-gatherers affect the evolution of human social behaviors? Science 324: 1293–1298.

Bowman DMJS, Balch JK, Artaxo P et al. 2009. Fire in the Earth system. Science 324: 381–384.

Boyce CK, Brodribb TJ, Feild TS, and Zwieniecki MA. 2009. Angiosperm leaf vein evolution was physiologically and environmentally transformative. Proc. Roy. Soc. B 276: 1771–1776.

Boyce CK, Hotton CL, Fogel ML et al. 2007. Devonian landscape heterogeneity recorded by a giant fungus. Geology 35: 399–402.

Boyce CK and Knoll AH. 2002. Evolution of developmental potential and the multiple independent origins of leaves in Paleozoic vascular plants. Paleobiology 28: 70–100.

Boyce CK and Lee J-E. 2010. An exceptional role for flowering plant physiology in the expansion of tropical rain forests and biodiversity. Proc. Roy. Soc. B 277: 3437–3443.

Boyce CK and Lee J-E. 2011. Could land plant evolution have fed the marine revolution? Paleont. Res. 15: 100–105.

Boyce CK and Leslie AB. 2012. The paleontological context of angiosperm vegetative evolution. Intern. J. Plant Sci. 173: 561–568.

Boyce CK and Zwieniecki MA. 2019. The prospects for constraining productivity through time with the whole-plant physiology of fossils. New Phytologist 223: 40–49.

Bradbury RH and Young PC. 1982. The race and the swift revisited; or is aggression between corals important? Proc. Fourth Intern. Coral Reef Symp. 2: 351–356.

Bramble DM and Lieberman DE. 2004. Endurance running and the evolution of *Homo*. Nature 432: 345–352.

Brandon R and McShea D. 2020. The missing two-thirds of evolutionary theory. Cambridge University Press, Cambridge.

Brazeau MD and Friedman M. 2015. The origin and early phylogenetic history of jawed vertebrates. Nature 520: 490–497.

Brocks JJ, Jarrett AJM, Sirantoine E et al. 2017. The rise of algae in the Cryogenian oceans and the emergence of animals. Nature 548: 578–581.

Brodie ED Jr, Ridenhour BJ, and Brodie ED III. 2002. The evolutionary response of predators to dangerous prey: Hotspots and coldspots in the geographic mosaic of coevolution between garter snakes and newts. Evolution 57: 2067–2082.

Brodribb TJ, Jordan GJ, and Carpenter RJ. 2013. Unified changes in cell size permits coordinated leaf evolution. New Phytologist 199: 559–570.

Brodribb TJ and McAdam SAM. 2011. Passive origins of stomatal control in vascular plants. Science 331: 582–585.

BroJørgensen J. 2013. Evolution of sprint speed in African savanna herbivores in relation to predation. Evolution 67: 3371–3376.

Brosse M, Bucher H, Baud A et al. 2019. New data from Oman indicate benthic high biomass productivity coupled with low taxonomic diversity in the aftermath of the Permian-Triassic boundary mass extinction. Lethaia 52: 165–187.

Brown JH, Burnside WR, Davidson AD et al. 2011. Energetic limits to economic growth. BioScience 61: 19–26.

Brown JH, Gillooly JF, Allen AP et al. 2004. Toward a metabolic theory of ecology. Ecology 85: 1771–1789.

Brunet M and Jaeger J-J. 2017. From the origin of the anthropoids to the first appearance of the human family. C.R. Palevol. 16: 189–195.

Budd GE and Jensen S. 2017. The origin of the animals and a "Savannah" hypothesis for early bilaterian evolution. Biol. Revs. 92: 446–473.

Bullen CD, Campos AA, Grego EJ et al. 2021. The ghost of a giant—six hypotheses for how an extinct megaherbivore structured kelp forests across the North Pacific rim. Glob. Ecol. Biogeogr. 30: 2101–2118.

Burness GP, Diamond J, and Flannery T. 2001. Dinosaurs, dragons, and dwarfs: The evolution of maximum body size. Proc. Nat. Acad. Sci. U.S.A. 98: 14518–14523.

Burrows M. 2014. Jumping mechanisms in dictyopharid planthoppers (Hemiptera, Dictyopharidae). J. Exp. Biol. 217: 402–413.

Bush AM, Hunt G, and Bambach RK. 2016. Sex and the shifting biodiversity dynamics of marine animals in deep time. Proc. Nat. Acad. Sci. U.S.A. 113: 14073–14078.

Buss LW. 1987. The evolution of individuality. Princeton University Press, Princeton.

Butterfield NJ. 2009. Modes of pre-Ediacaran multicellularity. Precambrian Res. 173: 201–211.

Butterfield NJ. 2015. Proterozoic photosynthesis—a critical review. Palaeontology 58: 953–972.

Butterfield NJ. 2018. Oxygen, animals and aquatic bioturbation: an updated account. Geobiology 16: 3–16.

Butterfield NJ. 2000. *Bangiomorpha pubescens* n. gen., n sp.: implications for the evolution of sex, multicellularity, and the Mesoproterozoic-Neoproterozoic radiation of eukaryotes. Paleobiology 26: 386–404.

Butterfield NJ. 2022. Constructional and functional anatomy of Ediacaran rangeomorphs. Geol. Mag. 159: 1148–1159.

Cairns DK, Gaston AJ, and Huettmann F. 2008. Endothermy, ectothermy and the global structure of marine vertebrate communities. Mar. Ecol. Prog. Ser. 356: 239–250.

Canfield DE. 2021. Carbon cycle evolution before and after the Great Oxidation of the atmosphere. Amer. J. Sci. 321: 297–331.

Canfield DE, Glazer AN, and Falkowski PG. 2010. The evolution and future of Earth's nitrogen cycle. Science 330: 192–196.

Canfield DE, Rosing MT, and Bjerrum C. 2006. Early anaerobic metabolisms. Phil. Trans. Roy. Soc. 363: 1819–1836.

Cannell AERO. 2018. The engineering of the giant dragonflies of the Permian: Revised body mass, power, air supply, thermoregulation and the role of air density. J. Exp. Biol. 221: jeb185405.

Carballido JL, Pol D, Otero A et al. 2017. A new giant titanosaur sheds light on body mass evolution among sauropod dinosaurs. Proc. Roy. Soc. B 284: 2017.1219.

Carbone C, Maddox T, Funston PJ et al. 2009. Parallels between playbacks and Pleistocene tar seeps suggest sociality in an extinct sabretooth cat. Smilodon. Biol. Lett. 5: 81–85.

Carrillo JD, Faurby S, Silvestro D et al. 2020. Disproportionate extinction of South American mammals drove the asymmetry of the Great American Biotic Interchange. Proc. Nat. Acad. Sci. U.S.A. 117: 26281–26287.

Carriquí M, Roig-Oliver M, Brodribb TJ et al. 2019. Anatomical constraints to nonstomatal diffusion conductance and photosynthesis in lycophytes and bryophytes. New Phytologist 222: 1256–1270.

Carroll SB. 2005. Endless forms most beautiful: The new science of evo devo and the making of the animal kingdom. W. W. Norton, New York.

Carroll SM. 2010. From eternity to here: The quest for the ultimate theory of time. Dutton/Penguin, New York.

Cavalier-Smith T. 2017. Origin of animal multicellularity: Precursors, causes, consequences—the choanoflagellate/sponge transition, neurogenesis and the Cambrian explosion. Phil. Trans. Roy. Soc. 372: 2015.0476.

Centeno-González NK, Martínez-Cabrera HI, Porras-Múzquiz H, and Estrada-Ruiz E. 2021. Late Campanian fossil of a legume fruit supports Mexico as a center of Fabaceae radiation. Commun. Biol. 4: 41.

Chaisson E. 2005. Epic of evolution: Seven ages of the cosmos. Columbia University Press, New York.

Chamberlain JA Jr. 1991. Cephalopod locomotor design and evolution: The constraints of jet propulsion. In Rayner JMV and Wootton RJ (eds.), Biomechanics in evolution. Cambridge University Press, Cambridge: 57–98.

Chan BKK, Dreyer N, Gale AS et al. 2021. The evolutionary diversity of barnacles, with an updated classification of fossil and living forms. Zool. J. Linn. Soc. 193: 789–847.

Chase MW. 2004. Monocot relationships: An overview. Amer. J. Bot. 1645–1655.

Chatterjee S, Templin RJ, and Campbell KE. 2007. The aerodynamics of *Argentavis*, the world's largest flying bird from the Miocene of Argentina. Proc. Nat. Acad. Sci. U.S.A. 104: 12398–12404.

Checa AG and Jiménez-Jiménez AP. 1998. Constructional morphology, origin, and the evolution of the gastropod operculum. Paleobiology 24: 109–132.

Cheng JT, Tracy JL, Foulsham T et al. 2013. Two ways to the top: Evidence that dominance and prestige are distinct yet viable avenues to social rank and influence. J. Personality and Social Psychol. 104: 103–125.

Chiarenza AA, Mannion FD, Farnsworth A et al. 2022. Climatic constraints on the biogeographic history of Mesozoic dinosaurs. Current Biology 32: 570–585.

Childress JJ and Girguis PR. 2011. The metabolic demands of endosymbiotic chemoautotrophic metabolism on host physiological capacities. J. Exp. Biol. 214: 312–325.

Chua A. 2007. Day of empire: How hyperpowers rise to global dominance—and why they fall. Doubleday, New York.

Clark G. 2007. A farewell to alms: A brief economic history of the world. Princeton University Press, Princeton.

Clark JW, Harris BJ, Hetherington AJ et al. 2022. The origin and evolution of stomata. Current Biology 32: R539-R553.

Clarke A. 2013. Dinosaur energetics: Setting the bounds on feasible physiologies and ecologies. Amer. Nat. 182: 283–297.

Cloyed CS, Grady JM, Savage VM et al. 2021. The allometry of locomotion. Ecology 102: e03369.

Clutton-Brock T. 2007. Sexual selection in males and females. Science 318: 1882–1885.

Coatham SJ, Vinther J, Rayfield EJ, and Klug C. 2020. Was the Devonian placoderm *Titanichthys* a suspension feeder? Roy. Soc. Open Sci. 7: 200272.

Coen E. 2012. Cells to civilizations: The principles of change that shape life. Princeton University Press, Princeton.

Cole DB, Mills DB, Erwin DH et al. 2020. On the co-evolution of surface oxygen levels and animals. Geobiology 18: 260–281.

Collinson ME and Hooker JJ. 2000. Gnaw marks on Eocene seeds: Evidence for early rodent behaviour. Palaeogeogr., Palaeoclimatol., Palaeoecol. 157: 127–149.

Collopy MW. 1983. Foraging behavior and success of golden eagles. Auk 100: 747–749.

Conway Morris S. 2003. Life's solution: Inevitable humans in a lonely universe. Cambridge University Press, Cambridge.

Cooper JA, Pimiento C, Ferrón HG, and Benton MJ. 2020. Body dimensions of the extinct giant shark *Otodus megalodon*: A 2D reconstruction. Sci. Reps. 10: 14596.

Corning PA. 2005. Holistic Darwinism: Synergy, cybernetics, and the bioeconomics of evolution. University of Chicago Press, Chicago.

Cowen R. 1970. Analogies between the recent bivalve *Tridacna* and the fossil brachiopods Lyttoniacea and Richthofeniacea. Palaeogeogr., Palaeocllmatol, Palaeoecol. 8: 329–344.

Cowen R. 1981. Crinoid arms and banana plantations: An economic harvesting analogy. Paleobiology 7: 332–343.

Cowen R. 1983. Algal symbiosis and its recognition in the fossil record. In Tevesz MJS and McCall PL (eds.), Biotic interactions in recent and fossil benthic communities. Plenum, New York: 431–478.

Coyne JA. 2009. Why evolution is true. Viking, New York.

Crampton, WGR. 2019. Electoreception, electrogenesis and electric signal evolution. J. Fish Biol. 95: 92–134.

Cristoffer C and Peres CA. 2003. Elephants versus butterflies: The ecological role of herbivores in the evolutionary history of two tropical worlds J. Biogeogr. 30: 1357–1380.

Crofoot MC and Wrangham RW. 2010. Intergroup aggression in primates and humans: The case for a unified theory. In Kappeler PM and Silk JB (eds.), Mind the gap: Tracing the origins of human universals. Springer, Berlin: 171–195.

Croft DA and Lorente M. 2021. No evidence for parallel evolution of cursorial limb adaptations among Neogene South American native ungulates (SANU's). PLoS One 16: e0256371.

Crowe SA, Dössing LN, Beukes NJ et al. 2013. Atmospheric oxygenation three billion years ago. Nature 501: 535–539.

Cyr H and Pace ML. 1993. Magnitude and patterns of herbivory in aquatic and terrestrial ecosystems. Nature 361: 147–150.

Czech B. 2019. The trophic theory of money: Principles, corollaries, and policy implications. J. Proc. Roy. Soc. New South Wales 152: 66–81.

Dai T, Zhang X-L, Peng S-C, and Yang B. 2019. Enrolment and trunk segmentation of a Cambrian eodiscoid trilobite. Lethaia 52: 502–512.

Daley AC and Bergström J. 2012. The oral cone of *Anomalocaris* is not a classic "Peytoia." Naturwissenschaften 99: 501–504.

Daley AC, Paterson JR, Edgecombe GD et al. 2013. New anatomical information on *Anomalocaris* from the Cambrian Emu Bay Shale of South Australia and a reassessment of its inferred predatory habits. Palaeontology 56: 971–990.

D'Antonio MP and Boyce CK. 2020. Arborescent lycopsid periderm production was limited. New Phytologist 228: 741–751.

Darimont CT, Carlson SM, Kinnison MJ et al. 2009. Human predators outpace other agents of trait change in the wild. Proc. Nat. Acad. Sci. U.S.A. 106: 952–954.

Darimont CT, Fox CH, Bryan HM, and Reimchen TE. 2015. The unique ecology of human predators. Science 349: 858–860.

Darwin J. 2007. After Tamerlane: The rise and fall of global empires, 1400–2000. Penguin, New York.

Dasilao JC Jr and Sasaki K. 1998. Phylogeny of the flyingfish family Exocoetidae (Teleostei, Beloniformes). Ichthyol. Res. 45: 347–353.

Dawkins R and Krebs JR. 1979. Arms races between and within species. Proc. Roy. Soc. London B 205: 489–511.

Dean JM and Smith AP. 1978. Behavioral and morphological adaptations of a tropical plant to high rainfall. Biotropica 10: 152–154.

DeAngelis DL. 1992. Dynamics of nutrient cycling and food webs. Chapman and Hall, London.

Dejean A, Solano PJ, Ayroles J et al. 2005. Arboreal ants build traps to capture prey. Nature 434: 973.

Del Cortona A, Jackson CJ, Bucchini F et al. 2020. Neoproterozoic origin and multiple transitions to macroscopic growth in green seaweeds. Proc. Nat. Acad. Sci. U.S.A. 117: 2551–2559.

DeLong JP. 2008. Maximum power principle predicts the outcomes of two-species competition experiments. Oikos 117: 1329–1336.

Dennett DC. 2006. Breaking the spell: Religion as a natural phenomenon. Penguin, New York.

Denny M. 1988. Biology and the mechanics of the wave-swept environment. Princeton University Press, Princeton.

Denny M. 1993. Air and water: The biology and physics of life's media. Princeton University Press, Princeton.

Derry LA, Kurtz AC, Ziegler K, and Chadwick OA. 2005. Biological control of terrestrial silica cycling and export fluxes to watersheds. Nature 433: 728–731.

DeVore PR, Nyandwi A, Eckardt W et al. 2020. Urticaceae leaves with stinging trichomes were already present in latest Early Eocene Okanogan Highlands, British Columbia, Canada. Amer. J. Bot. 107: 1449–1456.

DeVries MS, Murphy EAR, and Patek SN. 2012. Strike mechanics of an ambush predator: The spearing mantis shrimp. J. Exp. Biol. 215: 4374–4384.

D'Hondt S, Donaghay P, Zachos JC et al. 1998. Organic carbon fluxes and ecological recovery from the Cretaceous-Tertiary mass extinction. Science 282: 276–279.

Diamond J. 1997. Guns, germs, and steel: The fate of human societies. W. W. Norton, New York.

Dietl GP. 2003. Coevolution of a marine gastropod predator and its dangerous bivalve prey. Biol. J. Linn. Soc. 80: 409–436.

Dietl GP. 2004. Origins and circumstances of adaptive divergence in whelk feeding behavior. Palaeogeogr., Palaeoclimatol., Palaeoecol. 208: 279–291.

Dijk TA van. 1989. Structures of discourse and structures of power. Ann. Intern. Communications Assoc. 12: 18–59.

Ding W, Dong L, Sun Y et al. 2019. Early animal evolution and highly oxygenated seafloor niches hosted by microbial mats. Sci. Reps. 9: 13628.

Donley JM, Sepulveda CA, Konstantinidis M et al. 2004. Convergent evolution in mechanical design of lamnid sharks and tunas. Nature 429: 61–65.

Dorrington GE. 2016. Heavily loaded flight and limits to the maximum size of dragonflies (Anisoptera) and griffenflies (Meganisoptera). Lethaia 49: 269–274.

Doughty CE. 2017. Herbivores increase the global availability of nutrients over millions of years. Nature Ecol. Evol. 1: 1820–1827.

Dreisig H. 1981. The rate of predation and its temperature dependence in a tiger beetle, *Cicindela hybrida*. Oikos 36: 196–202.

Dressaire E, Yamada L, Song B, and Roper M. 2016. Mushrooms use convectively created airflows to disperse their spores. Proc. Nat. Acad. Sci. U.S.A. 113: 2833–2838.

Droser ML, Jensen S, and Gehling JG. 2002. Trace fossils and substrates of the terminal Proterozoic-Cambrian transition: Implications for the record of early bilaterians and sediment mixing. Proc. Nat. Acad. Sci. U.S.A. 99: 12572–12576.

Droser ML, Tarhan LG, and Gehling JG. 2017. The rise of animals in a changing environment: Global ecological innovation in the Late Ediacaran. Ann. Rev. Earth Planet. Sci. 45: 593–617.

Duarte CM. 1995. Submerged aquatic vegetation in relation to different nutrient regimes. Ophelia 41: 87–112.

Dudley R and DeVries P. 1990. Tropical rain forest structure and the geographical distribution of gliding vertebrates. Biotropica 22: 432–434.

Dudley R and Vermeij GJ. 1994. Energetic constraints of folivory: Leaf fractionation by frugivorous bats. Func. Ecol. 8: 668.

Duve C de. 2005. Singularities: Landmarks on the pathways of life. Cambridge University Press, Cambridge.

Dvořáček J, Senhadová H, Weyda F et al. 2020. First comprehensive study of a giant among the insects, *Titanus giganteus*: Basic facts from its biochemistry, physiology, and anatomy. Insects 11: 120.

Dzik J. 2005. Behavioral and anatomical unity of the earliest burrowing animals and the cause of the "Cambrian explosion." Paleobiology 31: 503–521.

Edwards D and Axe L. 2012. Evidence for a fungal affinity for *Nemataschetum*, a close ally of *Prototaxites*. Bot. J. Linn. Soc. 168: 1–18.

Eichenseer K, Balthasar U, Smart CW et al. 2019. Jurassic shift from abiotic to biotic control on marine ecological success. Nature Geosci. 12: 635–642.

Emlen DJ. 2008. The evolution of animal weapons. Ann. Rev. Ecol. Syst. 39: 387–413.

Emmons LH. 1989. Jaguar predation on chelonians. J. Herpetol. 23: 311–314.

Epihov DZ, Batterman SA, Hedin LO et al. 2017. N_2-fixing tropical legume evolution: A contributor to enhanced weathering through the Cenozoic. Proc. Roy. Soc. B 284: 2017.0370.

Epihov DZ, Saltonstall K, Batterman SA et al. 2021. Legume-microbiome interactions unlock mineral nutrients in regrowing tropical forests. Proc. Nat. Acad. Sci. U.S.A. 118: e20222411118.

Erickson GM, Krick BA, Hamilton M et al. 2012a. Complex dental structure and wear biomechanics in hadrosaurid dinosaurs. Science 338: 98–101.

Erickson GM, Gignac PM, Steppan SJ et al. 2012b. Insights into the ecology and evolutionary success of crocodilians revealed through bite-force and tooth-pressure experimentation. PLoS One 7: e31781.

Eriksson ME and Horn E. 2017. *Agnostus pisiformis*—a half a billion-year old pea-shaped enigma. Earth-Sci. Revs. 173: 65–76.

Erwin DH. 1996. Understanding biotic recoveries: Extinction, survival, and preservation during the end-Permian mass extinction. In Jablonski D, Erwin DH, and Lipps JH (eds.), Evolutionary paleobiology: In honor of James W. Valentine. University of Chicago Press, Chicago: 398–418.

Estes JA, Burdin A, and Doak DF. 2016. Sea otters, kelp forests, and the extinction of Steller's sea cow. Proc. Nat. Acad. Sci. U.S.A. 113: 880–885.

Estes JA and Palmisano JF. 1974. Sea otters: Their role in structuring nearshore communities. Science 185: 1058–1060.

Estes JA, Terborgh J, Brashnes JS et al. 2011. Trophic downgrading of planet Earth. Science 333: 331–336.

Evans SD, Gehling JG, and Droser ML. 2019. Slime travelers: Early evidence of animal mobility and feeding in an organic mat world. Geobiology 17: 490–509.

Ezcurra MD, Nesbitt SJ, Bronzatti M et al. 2020. Enigmatic dinosaur precursors bridge the gap to the origin of Pterosauria. Nature 588: 445–449.

Falkowski PG, Katz ME, Knoll AH et al. 2004. The evolution of modern eukaryotic phytoplankton. Science 305: 354–360.

Faurby S, Silvestro D, Werdelin L, and Antonelli A. 2020. Brain expansion in early hominins predicts carnivore extinctions in East Africa. Ecol. Lett. 23: 537–544.

Fenberg PB and Roy K. 2008. Ecological and evolutionary consequences of size-selective harvesting: How much do we know? Molec. Ecol. 17: 209–220.

Ferguson N. 2008. The ascent of money: A financial history of the world. Penguin, New York.

Ferretti MP. 2007. Evolution of bone-cracking adaptations in hyaenids (Mammalia, Carnivora). Swiss J. Geosci. 100: 41–52.

Ferrón H. 2017. Regional endothermy as a trigger for gigantism in some extinct macropredatory sharks. PLoS One 12: e0185185.

Ferrón HG, Martínez-Pérez C, and Botella H. 2018. The evolution of gigantism in active marine predators. Hist. Biol. 30: 712–716.

Finnegan S, McClain C, Kosnik MA, and Payne JL. 2011. Escargots through time: An energetic comparison of marine gastropod assemblages before and after the Mesozoic Marine Revolution. Paleobiology 37: 252–269.

Fix B. 2018. Hierarchy and the power-law income distribution tail. J. Comput. Soc. Sci. 1: 471–491.

Fix B. 2019. Energy, hierarchy and the origin of inequality. PLoS One 14: e0215692.

Fleischle CV, Wintrich T, and Sander PM. 2018. Quantitative histological models suggest endothermy in plesiosaurs. PeerJ 6: e4955.

Fournier GP, Moore KR, Rangel LT et al. 2021. Archean origin of oxygenic photosynthesis and extant cyanobacterial lineages. Proc. Roy. Soc. B. 288: 2021.0675.

Fox CP, Whiteside JH, Olsen PE et al. 2022. Two-pronged kill mechanism at the end-Triassic mass extinction. Geology 50: 448–453.

Francis L. 1991. Sailing downwind: Aerodynamic performance of the Velella sail. J. Exp. Biol. 158: 117–132.

Fratzl P and Barth FG. 2009. Biomaterial systems of mechanosensing and actuation. Nature 462: 442–448.

Frazzetta TH and Kardong KV. 2002. Prey attack by a large theropod dinosaur. Nature 416: 387–388.

Freestone AL, Torchin ME, Jurgens LJ et al. 2021. Stronger predation intensity and impact on prey communities in the tropics. Ecology 102: e03428.

Frey E, Sues H-D, and Munk W. 1997. Gliding mechanism in the Late Permian reptile *Coelurosauravus*. Science 275: 1450–1452.

Friedman M, Shimada K, Eberhart MJ et al. 2013. Geographic and stratigraphic distribution of the Late Cretaceous suspension-feeding bony fish *Bonnerichthys gladius* (Teleostel, Pachycormiformis). J. Vert. Paleont. 33: 35–47.

Friedman M, Shimada K, Martin LD et al. 2010. 100-million-year dynasty of giant planktivorous bony fishes in the Mesozoic seas. Science 327: 990–993.

Friis EM, Crane PR, Pedersen KR et al. 2015. Exceptional preservation of tiny embryos documents seed dormancy in early angiosperms. Nature 528: 551–554.

Fritz J, Hummel J, Kienzle E et al. 2011. Gizzard vs. teeth, it's a tie: Food-processing efficiency in herbivorous birds and mammals and implications for dinosaur feeding strategies. Paleobiology 37: 577–586.

Fröbisch NB, Fröbisch J, Sander PM et al. 2013. Macropredatory ichthyosaur from the Middle Triassic and the origin of modern trophic networks. Proc. Nat. Acad. Sci. U.S.A. 110: 1393–1397.

Frouz J. 2018. Effects of soil macro- and mesofauna on litter decomposition and soil organic matter stabilization. Geoderma 332: 161–172.

Fry BG, Vidal N, Norman JA et al. 2006. Early evolution of the venom system in lizards and snakes. Nature 439: 584–588.

Fukuyama F. 2022. Liberalism and its discontents. Farrar, Straus and Giroux, New York.

Galbraith JK. 1983. The anatomy of power. Houghton Mifflin, New York.

Galofré-Vilà G, Meissner CM, McKee M, and Stuckler D. 2021. Austerity and the rise of the Nazi Party. J. Econ. Hist. 81: 81–113.

Galstyan A and Hay A. 2018. Snap, crack and pop of explosive fruit. Curr. Op. Genet. Devel. 51: 31–36.

Gans C and De Vree F. 1986. Shingle-back lizards crush snail shells using temporal summation (tetanus) to increase the force of the adductor muscles. Experientia 42: 387–389.

Garland T. Jr. 1983. The relation between maximal running speed and body mass in terrestrial mammals. J. Zool. London 199: 157–170.

Garvin J, Buick R, Anbar AD et al. 2009. Isotopic evidence for an aerobic nitrogen cycle in the latest Archean. Science 323: 1045–1048.

Geist V. 1978. On weapons, combat, and ecology. In Krames L, Pliner P, and Alloway T (eds.), Advances in the study of communication and effect, Vol. 4: Aggression, dominance and individual spacing. Plenum, New York: 1–30.

Geyer G, Pals MC, and Wotte T. 2020. Unexpectedly curved spines in a Cambrian trilobite: Considerations on the spinosity in *Kingaspidoides spinirecurvatus* sp. nov. from the Anti-Atlas, Morocco, and related Cambrian ellipsocephaloids. Paläont. Z. 94: 645–660.

Ghisalberti M, Gold DA, Laflamme M, Clapham ME et al. 2014. Canopy flow analysis reveals the advantage of size in the oldest communities of multicellular eukaryotes. Curr. Biol. 24: 1–5.

Gibiansky ML, Conrad JC, Jin F et al. 2010. Bacteria use type IV pili to walk upright and detach from surfaces. Science 330: 197.

Giezen M van der and Lenton TM. 2012. The rise of oxygen and complex life. J. Eukaryotic Microbiol. 59: 111–113.

Gignac PM and Erickson GM. 2017. The biomechanics behind extreme osteophagy in *Tyrannosaurus rex*. Sci. Reps. 7: 2012.

Gilbert PUPA, Bergmann KD, Boekelheide N et al. 2022. Biomineralization: Integrating metabolism and evolutionary history. Sci. Adv. 8: eab19653.

Gill FB. 1985. Hummingbird flight speeds. Auk. 102: 97–101.

Gill RE Jr, Tibbitts TL, Douglas DC et al. 2009. Extreme endurance flights by landbirds crossing the Pacific Ocean: Ecological corridor rather than barrier? Proc. Roy. Soc. B 276: 447–457.

Gillooly J F, Brown JH, West GB et al. 2001. Effects of size and temperature on metabolic rate. Science 293: 2248–2251.

Glasspool IJ, Scott AC, Waltham D et al. 2015. The impact of fire on the Late Paleozoic Earth system. Front. Plant Sci. 3: 00756.

Glazier DS. 2015. Is metabolic rate a universal "pacemaker" for biological processes? Biol. Revs. 90: 377–407.

Goeij JM de, Moodley L, Houtekamer M et al. 2008. Tracing ^{13}C-enriched dissolved and particulate organic carbon in the bacteria-containing coral reef sponge *Galisarcha caerulea*: Evidence for DOM feeding. Limnol. Oceanogr. 53: 1377–1386.

Goeij JM de, van Oevelen D, Vermeij MJA et al. 2013. Surviving in a marine desert: The sponge loop retains resources within coral reefs. Science 342: 108–110.

Goldbogen JA, Pyenson ND, and Shadwick RE. 2007. Big gulps require high drag for fin whale lunge feeding. Mar. Ecol. Prog. Ser. 349: 289–301.

Goldbogen JA, Cade DE, Wisniewska DM et al. 2019. Why whales are big but not bigger: Physiological drivers and ecological limits in the age of ocean giants. Science 366: 1367–1372.

Gould SJ. 1985. The paradox of the first tier: An agenda for paleobiology. Paleobiology 11: 2–12.

Gould SJ. 1989. Wonderful life: The Burgess Shale and the nature of history. W. W. Norton, New York.

Gould SJ. 2002. The structure of evolutionary theory. Belknap Press of Harvard University, Cambridge.

Gould SJ and Calloway CB. 1980. Clams and brachiopods—ships that pass in the night. Paleoblology 6: 383–396.

Grady JM, Enquist BJ, Dettweiler-Robinson E et al. 2014. Evidence for mesothermy in dinosaurs. Science 344: 1268–1272.

Grady JM, Maitner BS, Winter AS et al. 2019. Metabolic asymmetry and the global diversity of marine predators. Science 363: eaat4220.

Grasso DA, Giannetti D, Castracani C et al. 2020. Rolling away: A novel context-dependent escape behaviour discovered in ants. Sci. Reps. 10: 3784.

Green PA, Thompson FJ, and Cant MA. 2022. Fighting force and experience combine to determine contest success in a warlike mammal. Proc. Nat. Acad. Sci. U.S.A. 119: e2119176119.

Grigg G, Nowack J, Bicudo JEP et al. 2022. Whole-body endothermy: Ancient, homologous and widespread among the ancestors of mammals, birds and crocodylians. Biol. Revs. 97: 766–801.

Grinsted L, Schou MF, Pettepani V et al. 2020. Prey to predator body size ratio in the evolution of cooperative hunting—a social spider test case. Devel. Genes Evol. 230: 173–184.

Grosberg RK and Strathmann RR. 2007. The evolution of multicellularity: A minor major transition. Ann. Rev. Ecol. Evol. Syst. 38: 621–654.

Guerrero R, Pedros-Alió C, Esteve I et al. 1986. Predatory prokaryotes: Predation and primary consumption evolved in bacteria. Proc. Nat. Acad. Sci. U.S.A. 83: 2138–2142.

Gumsley AP, Chamberlain KR, Bleeker W et al. 2017. Timing and tempo of the Great Oxidation Event. Proc. Nat. Acad. Sci. U.S.A. 114: 1811–1816.

Hallam A. 1991. Why was there a delayed radiation after the end-Paleozoic extinctions? Hist. Biol. 5: 257–262.

Hammond TP. 2019. It (still) takes a nation: Why private charity will never replace the welfare state. Independent Review 23: 521–537.

Hao J, Knoll AH, Huang F et al. 2020. Cycling phosphorus on the Archean Earth: Part II. Phosphorus limitation on primary production in Archean ecosystems. Geochim. Cosmochim. Acta 180: 360–377.

Harari YN. 2015. Sapiens: A brief history of humankind. Harper/HarperCollins, New York.

Hardin G. 1968. The tragedy of the commons. Science 162: 1243–1248.

Harrison CJ and Morris JL. 2018. The origin and early evolution of vascular plant shoots and leaves. Phil. Trans. Roy. Soc. B 373: 2016.0496.

Harshey RM, Kawagishi I, Maddock J, and Kenney LJ. 2003. Function, diversity, and evolution of signal transduction in prokaryotes. Developmental Cell 4: 459–465.

Hasenfuss I. 2008. The evolutionary pathway to insect flight—a tentative reconstruction. Arthropod Syst. Phylog. 66: 19–35.

Heck KL Jr, Samsonova M, Poore AGB, and Hyndes GA. 2021. Global patterns in seagrass herbivory: Why, despite existing evidence, there are solid arguments in favor of latitudinal gradients in seagrass herbivory. Estuaries and Coasts 44: 481–490.

Hector DP. 1986. Cooperative hunting and its relationship to foraging success and prey size in an avian predator. Ethology 73: 247–257.

Heide T van der, Govers LL, de Fouw J et al. 2012. A three-stage symbiosis forms the foundation of seagrass ecosystems. Science 336: 1432–1434.

Heim NA, Knope ML, Schaal EK et al. 2015. Cope's Rule in the evolution of marine animals. Science 347: 867–870.

Heinrich B. 1993. The hot-blooded insects: Strategies and mechanisms of thermoregulation. Harvard University Press, Cambridge.

Henehan MJ, Ridgwell A, Thomas E et al. 2019. Ocean acidification and protracted Earth system recovery following the end-Cretaceous Chicxulub impact. Proc. Nat. Acad. Sci. U.S.A. 116: 22500–22504.

Herbert GS. 2018. Evidence for intense biotic interactions in the eastern Gulf of Mexico after a two million year hiatus: inferences from muricid edge drilling behaviour. J. Molluscan Stud. 84: 427–431.

Herbert GS, Whitenack LB, and McKnight JY. 2017. Behavioural versatility of the giant murex *Muricanthus fulvescens* (Sowerby, 1834) (Gastropoda: Muricidae) in interactions with difficult prey. J. Molluscan Stud. 82: 357–365.

Hertweck KL, Kinney MS, Stuart SA et al. 2015. Phylogenetics, divergence times and diversification from three genomic partitions in monocots. Bot. J. Linn. Soc. 178: 375–393.

Hetherington AJ and Dolan L. 2018. Stepwise and independent origins of roots among land plants. Nature 561: 235–238.

Hibbett D, Flanchette R, Kenrick P, and Mills B. 2016. Climate, decay, and the death of coal forests. Curr. Biol. 26: R563–R567.

Hildenbrand A, Austermann G, Fuchs D et al. 2021. A potential cephalopod from the Early Cambrian of eastern Newfoundland, Canada. Commun. Biol. 4: 388.

Hirt MR, Jetz W, Rall BC, and Brose U. 2017. A general scaling law reveals why the largest animals are not the fastest. Nature Ecol. Evom. 1: 1116–1122.

Hofhuls H, Moulton D, Lessinnes T et al. 2016. Morphological innovation drives explosive seed dispersal. Cell 166: 222–233.

Hölldobler B and Wilson EO. 1990. The ants. Belknap Press of Harvard University, Cambridge.

Hölldobler B and Wilson EO. 2009. The superorganism: The beauty, elegance, and strangeness of insect societies. W. W. Norton, New York.

Honegger R, Edwards D, and Axe L. 2013. The earliest records of internally stratified cyanobacterial and algal lichens from the Lower Devonian of the Welsh Borderland. New Phytologist 197: 264–275.

Houghton IA, Koseff JR, Monismith SG, and Dabiri JO. 2018. Vertically migrating swimmers generate aggregation-scale eddies in a stratified ocean. Nature 556: 497–500.

Houssaye A, Lindgren J, Pellegrini R et al. 2013. Microanatomical and histological features in the long bones of mosasaurine mosasaurs (Reptilia, Squamata)—implications for aquatic adaptation and growth rates. PLoS One 8: e76741.

Hu G, Liu KS, Horvitz N et al. 2016. Mass seasonal bioflows of high-flying insect migrants. Science 354: 1584–1587.

Hummel J, Gee CT, Südekum K-H et al. 2008. *In vitro* digestibility of fern and gymnosperm foliage: Implications for sauropod feeding ecology and diet selection. Proc. Roy. Soc. B 275: 1015–1021.

Hutchinson JR and Garcia M. 2002. *Tyrannosaurus* was not a fast runner. Nature 415: 1018–1021.

Ilton M, Bhamla MS, Ma X et al. 2018. The principles of cascading lower limits in small, fast biological and engineered systems. Science 369: eaaz1082.

Iosilevskii G and Weihs D. 2008. Speed limits on swimming in fishes and cetaceans. J. Roy. Soc. Interface 5: 329–338.

Itescu Y, Schwarz O, Meiri S, and Pafilis P. 2017. Intraspecific competition, not predation, drives lizard tail loss on islands. J. Anim. Ecol. 86: 66–74.

Iwamoto M, Horikawa C, Shikata M et al. 2014. Stinging hairs on the Japanese nettle *Urtica thunbergiana* have a defensive function against mammalian but not insect herbivores. Ecol. Res. 29: 455–462.

Jablonski D and Bottjer DJ. 1990. Onshore-offshore trends in marine invertebrate evolution. In Ross RM and Allmon WD (eds.), Causes of evolution: A paleontological perspective. University of Chicago Press, Chicago: 21–75.

Jablonski D, Sepkoski JJ Jr, Bottjer DJ, and Sheehan PM. 1983. Onshore-offshore patterns in the evolution of Phanerozoic shelf communities. Science 222: 1123–1125.

Jackson JBC. 1997. Reefs since Columbus. Coral Reefs 16: S23–S32.

Jackson T. 2017. Prosperity without growth: Foundations for the economy of tomorrow (second edition). Routledge, New York.

Jacobs DK. 1992. Shape, drag, and power in ammonoid swimming. Paleobiology 18: 203–220.

Janis CM and Wilhelm PB. 1993. Were there mammalian pursuit predators in the Tertiary? J. Mammal. Evol. 1: 103–125.

Janzen DH. 1974. Tropical blackwater rivers, animals, and mast fruiting in the Dipterocarpaceae. Biotropica 6: 69–103.

Janzen DH. 1985. On ecological fitting. Oikos 45: 308–310.

Johnson BR, Borowiec ML, Chiu JC et al. 2013. Phylogenomics resolves evolutionary relationships among ants, bees, and wasps. Curr. Biol. 23: 2052–2062.

Johnson KG, Jackson JBC, and Budd AF. 2008. Caribbean reef development was independent of coral diversity over 28 million years. Science 319: 1521–1523.

Jones J. 1999. Cooperative foraging in the mountain caracara in Peru. Wilson Bull. 111: 437–439.

Kanso EA, Lopes RM, Strickler JR et al. 2021. Teamwork in the viscous oceanic microscale. Proc. Nat. Acad. Sci. U.S.A. 118: e2018193118.

Kantor YI and Puillandre N. 2012. Evolution of the radular apparatus in Conoidea (Gastropoda: Neogastropoda) as inferred from a molecular phylogeny. Malacologia 55: 55–90.

Karban R, LoPresti E, Vermeij GJ, and Latta R. 2019. Unidirectional grass hairs usher insects away from meristems. Oecologia 189: 711–718.

Karp AS, Faith JT, Marlon JR, and Staver AC. 2021. Global response of fire activity to Late Quaternary grazer extinctions. Science 374: 1145–1148.

Katija K and Dabiri JO. 2009. A viscosity-enhanced mechanism for biogenic ocean mixing. Nature 460: 624–626.

Kauffman SA. 1993. The origins of order: Self-organization and selection in evolution. Oxford University Press, New York.

Kauffman SAO. 2000. Investigations. Oxford University Press, Oxford.

Kauffman SAO. 2008. Reinventing the sacred: A new view of science, reason, and religion. Basic Books, New York.

Kay RF. 1981. The nut-crackers—a new theory of the adaptations of the Ramapithecinae. Amer. J. Phys. Anthropol. 55: 141–151.

Kerfoot WC. 1978. Combat between predatory copepods and their prey: *Cyclops*, *Epischura*, and *Bosmina*. Limnol. Oceanogr. 23: 1089–1102.

Kerfoot WC (ed.). 1980. Evolution and ecology of zooplankton communities. University Press of New England, Hanover: 10–27.

Kerfoot WC, Kellogg DL Jr, and Strickler JR. 1980. Visual observations of live zooplankters: Evasion, escape, and chemical defenses. In Kerfoot (ed.), Evolution and ecology of zooplankton communities.

Kidder DL and Erwin DH. 2001. Secular distribution of biogenic silica through the Phanerozoic: Comparison of silica-replaced fossils and bedded cherts at the series level. J. Geol. 109: 509–522.

Kidwell SM and Brenchley PJ. 1996. Evolution of the fossil record: thickness trends in marine skeletal accumulations and their implications. In Jablonski D, Erwin JH, and Lipps JH (eds.), Evolutionary paleobiology: in honor of James W. Valentine. University of Chicago Press, Chicago: 299–336.

Kiers ET and Denison RF. 2008. Sanctions, cooperation, and the stability of plant-rhizosphere mutualisms. Ann. Rev. Ecol. Evol. Syst. 39: 215–236.

Kiers ET, Duhamel M, Beesetty Y et al. 2011. Reciprocal rewards stabilize cooperation in the mycorrhizal symbiosis. Science 333: 880–882.

Kiers ET and West SAO 2015. Evolving new organisms via symbiosis. Science 348: 392–394.

Kiltie RA. 1982. Bite force as a basis for differentiation between rain forest peccaries (*Tayassu tajacu* and *T. pecari*). Biotropica 14: 188–195.

Klug C, Kröger B, Kiessling W et al. 2010. The Devonian nekton revolution. Lethaia 43: 465–477.

Knaus PL, van Heteren AH, Lungmus JK, and Sander PM. 2021. High blood flow into the femur indicates elevated aerobic capacity in synapsids since the Synapsida-Sauropsida split. Front. Ecol. Evol. 9: 451238.

Knoll AH and Bambach RK. 2000. Directionality in the history of life: diffusion from the left wall or repeated scaling of the right? Paleobiology 26: 1–14.

Knoll AH, Bambach RK, Canfield DE, and Grotzinger JP. 1996. Comparative Earth history and Late Permian mass extinction. Science 273: 452–457.

Knoll AH, Bambach RK, Payne JL et al. 2007. Paleophysiology and end-Permian mass extinction. Earth Planet. Sci. Lett. 256: 295–313.

Knoll AH and Follows MJ. 2016. A bottom-up perspective on ecosystem change in Mesozoic oceans. Proc. Roy. Soc. B 283: 2016.1755.

Koehl MAR and Alberte RS. 1988. Flow, flapping, and photosynthesis of *Nereocystis luetkeana*: a functional comparison of undulate and flat blade morphologies. Mar. Biol. 99: 435–444.

Koenen EJM, Ojeda DI, Bakker FT et al. 2021. The origin of legumes in a complex paleopolyploid phylogenomic tangle closely associated with the Cretaceous-Paleogene (K-Pg) mass extinction event. Syst. Biol. 70: 508–526.

Kohn AJ. 1956. Piscivorous gastropods of the genus Conus. Proc. Nat. Acad. Sci. U.S.A. 42: 168–171.

Kolmann MA, Welch MC, Summers AP, and Lovejoy NR. 2016. Always chew your food: freshwater stingrays use mastication to process tough insect prey. Proc. Roy. Soc. B 283: 2016.1392.

Kott A, Gart S, and Pusey J. 2021. From cockroaches to tanks: the same power-mass-speed relation describes both biological and artificial ground-mobile systems. PLoS One 16: e0259966.

Kozlov MV, Filippov BY, Zubrij NA, and Zverev V. 2015. Abrupt changes in invertebrate herbivory on woody plants at the forest-tundra ecotone. Polar Biol. 38: 967–974.

Kremer M. 1993. Population growth and technological change: one million B.C. to 1990. Quart. J. Econ. 108: 681–716.

Kruuk H. 1972. The spotted hyena: a study of predation and social behavior. University of Chicago Press, Chicago.

Kurland CG, Collins LJ, and Penny D. 2006. Genomics and the irreducible nature of eukaryotic cells. Science 312: 1011–1014.

Laakso TA, Strauss JV, and Peterson KJ. 2020. Herbivory and its effect on Phanerozoic oxygen concentrations. Geology 48: 410–414.

Labandeira CC. 2013. Deep-time patterns of tissue consumption by terrestrial arthropod herbivores. Naturwissenschaften 100: 355–364.

Labandeira CC, Tremblay SL, Bartowski KE, and VanAller Hernick L. 2014. Middle Devonian liverwort herbivory and antiherbivore defense. New Phytologist 202: 247–258.

LaBarbera M. 1983. The wheels won't go. Amer. Nat. 121: 395–408.

LaBarbera M. 1984. Feeding currents and particle capture mechanisms in suspension feeding animals. Amer. Zool. 24: 71–84.

Laflamme M, Xiao S, and Kowalewski M. 2009. Osmotrophy in modular Ediacara organisms. Proc. Nat. Acad. Sci. U.S.A. 106: 14438–14443.

Lambert O, Bianucci G, Post K et al. 2010. The giant bite of a new raptorial sperm whale from the Miocene epoch of Peru. Nature 466: 105–108 and 134.

Landes DS. 1998. The wealth and poverty of nations: Why some are so rich and some so poor. W. W. Norton, New York.

Lane N. 2009. Life ascending: The ten great inventions of evolution. W. W. Norton, New York.

Lane N. 2015. The vital question: Energy, evolution, and the origins of complex life. W. W. Norton, New York.

Lane N. and Martin W. 2010. The energetics of genome complexity. Nature 467: 929–934.

Lang J. 1971. Interspecific aggression by scleractinian corals. I The rediscovery of *Scolymia cubensis* (Milne Edwards and Haime). Bull. Mar. Sci. 21: 952–959.

Lang J. 1973. Interspecific aggression by scleractinian corals. II Why the race is not only to the swift. Bull. Mar. Sci. 23: 260–279.

Larramendi A. 2016. Shoulder height, body mass, and shape of proboscideans. Acta Palaeont. Polon. 61: 537–574.

Lavery TJ, Roudnew B, Seymour J et al. 2014. Whales sustain fisheries: Blue whales stimulate primary production in the Southern Ocean. Marine Mammal Sci. 30: 888–904.

Leigh EG, O'Dea A, and Vermeij GJ. 2014. Historical biogeography of the Isthmus of Panama. Biol. Revs. 89: 148–172.

Leigh EG Jr. 1983. When does the good of the group override the advantage of the individual? Proc. Nat. Acad. Sci. U.S.A. 80: 2985–2989.

Leigh EG Jr. 1999. Tropical forest ecology: A view from Barro Colorado Island. Oxford University Press, New York.

Leigh EG Jr. 2010. The evolution of mutualism. J. Evol. Biol. 23: 2507–2528.

Leigh EG Jr, Vermeij GJ, and Wikelski M. 2009. What do human economies, large islands and forest fragments reveal about the factors limiting ecosystem evolution? J. Evol. Biol. 22: 1–12.

Leigh EG Jr. and Ziegler C. 2019. Nature strange and beautiful: How living beings evolved and made the Earth a home. Yale University Press, New Haven.

Lenton TM. 1998. Gaia and natural selection. Nature 394: 439–447.

Levin SA. 1999. Fragile dominion: Complexity and the commons. Perseus Books, Reading.

Lev-Yadun S. 2001. Aposematic (warning) coloration associated with thorns in higher plants. J. Theoret. Biol. 210: 385–388.

Lev-Yadun S. 2014. Defensive masquerade by plants. Biol. J. Linn. Soc. 113: 1162–1166.

Lev-Yadun S. 2016. Does the whistling thorn acacia (*Acacia drepanolobium*) use auditory aposematism to deter mammalian herbivores? Plant Signaling & Behavior 11: e1207035.

Lewallen EA, Pitman RL, Kjartanson SL, and Lovejoy NR. 2011. Molecular systematics of flying fishes (Teleostei: Exocoetidae): Evolution in the epipelagic zone. Biol. J. Linn. Soc. 102: 161–174.

Lewis, WA. 1955. The theory of economic growth. Richard D. Erwin, Homewood.

Li H, Sun C-Y, Fang Y et al. 2020. Biomineral armor in leaf-cutter ants. Nature Comm. 11: 5792.

Lieberman P. 2007. The evolution of human speech: Its anatomical and neural bases. Curr. Anthropol. 48: 39–66.

Liechti F, Witvliet W, Weber R, and Bächler E. 2013. First evidence of a 200-day non-stop flight in a bird. Nature Comm. 4: 2554.

Ligrone R, Duckett JG, and Renzaglia KS. 2012. Major transitions in the evolution of early land plants: A bryological perspective. Ann. Bot. 109: 851–871.

Lindert PH. 2004. Growing public: Social spending and economic growth since the eighteenth century. Vol. I: The story. Cambridge University Press, Cambridge.

Lindgren J, Sjövall P, Thiel V et al. 2018. Soft-tissue evidence of homeothermy and crypsis in a Jurassic ichthyosaur. Nature 564: 359–365.

Liston J, Newbrey MG, Challands TJ, and Adams CE. 2013. Growth, age and size of the Jurassic pachycormid *Leedsichthys problematicus* (Osteichthyes: Actinopterygia). In Arratia G, Schultze H-P, and Wilson MVH (eds.), Mesozoic fishes 5—global diversity and evolution. Dr. Friedrich Pfeil, Munchen: 145–175.

Little AG, Bougheed SC, and Moyes CD. 2010. Evolutionary affinity of billfishes (Xiphiidae and Istiophoridae) and flatfishes (Pleuronectifores): Independent and trans-subordinal origins of endothermy in teleost fishes. Molec. Phylog. Evol. 56: 897–904.

Littler PM, Littler DS, and Brooks BL. 2005. Extraordinary mound building Avrainvillea (Chlorophyta): The largest tropical marine plants. Coral Reefs 24: 555.

Liu AG, McIlroy D, and Brasier 10. 2010. First evidence for locomotion in the Ediacara biota from the 565 Ma Mistaken Point Formation, Newfoundland. Geology 38: 123–126.

Liu C, Fu D, and Zhang X. 2021. Phosphatic carapace of the waptiid arthropod *Chuandianella ovata* and biomineralization in ecdysozoans. Palaeontology 64: 755–763.

LoDuca ST, Bykova N, Wu M et al. 2017. Seaweed morphology and ecology during the great animal diversification events of the Early Paleozoic: A tale of two floras. Geobiology 15: 588–616.

Longo SJ, Ray W, Farley GM et al. 2021. Snaps of a tiny amphipod push the boundary of ultrafast, repeatable movement. Curr. Biol. 31: R116–R117.

Losos JB. 2017. Improbable destinies: Fate, chance, and the future of evolution. Riverhead Books, New York.

Lotka AJ. 1922. Contribution to the energetics of evolution. Proc. Nat. Acad. Sci. U.S.A. 8: 147–151.

Lovegrove BG. 2017. A phenology of the evolution of endothermy in birds and mammals. Biol. Revs. 92: 1213–1240.

Lovvorn JR. 2010. Modeling profitability for the smallest marine endotherm: Auklets foraging within pelagic prey patches. Aquat. Biol. 8: 203–219.

Lowery CM, Bralower TJ, Owens JD et al. 2018. Rapid recovery of life at ground zero of the end-Cretaceous mass extinction. Nature 558: 288–291.

Lynch M. 2007. The frailty of adaptive hypotheses for the origins of organismal complexity. Proc. Nat. Acad. Sci. U.S.A. 104: 8597–8604.

Lyson TR, Miller IM, Bercovici AD et al. 2019. Exceptional continental record of biotic recovery after the Cretaceous-Paleogene mass extinction. Science 366: eaay2268.

MacCulloch D. 2010. Christianity: The first three thousand years. Viking, New York.

Mailho-Fontana PL, Antoniazzi Alexandre C et al. 2020. Morphological evidence of an oral venom system in cecilian amphibians. iScience 23: 101234.

Maisey JG, Bronson AW, Williams RR, and McKenzie M. 2017. A Pennsylvanian "supershark" from Texas. J. Vert. Paleont. 37: e1325369.

Mann KG. 1973. Seaweeds: Their productivity and strategy of growth. Science 182: 975–981.

Manos PS and Stanford AM. 2001. The historical biogeography of Fagaceae: Tracking the Tertiary history of temperate and subtropical forests of the northern hemisphere. Intern. J. Plant Sci. 162: S77–S93.

Manzuetti A, Perea D, Jones W et al. 2020. An extremely large saber tooth cat skull from Uruguay (Late Pleistocene-Early Holocene, Dolores Formation): Body size and paleobiological implications. Alcheringa 44: 332–339.

Marden JH and Allen RL. 2002. Molecules, muscles, and machines: Universal performance characteristics of motors. Proc. Nat. Acad. Sci. U.S.A. 99: 4161–4166.

Marden JH and Chai P. 1991. Aerial predation and butterfly design: How palatability, mimicry, and the need for evasive flight constrain mass allocation. Amer. Nat. 138: 15–36.

Marshall CD, Guzman A, Narazaki T et al. 2012. The ontogenetic scaling of bite force and head size in loggerhead sea turtles (*Caretta caretta*): Implications for durophagy in neritic, benthic habitats. J. Exp. Biol. 215: 4166–4174.

Martens EA, Wadhwa N, Jacobsen NS et al. 2015. Size structures sensory hierarchy in ocean life. Proc. Roy. Soc. B 282: 2015.1346.

Maynard Smith J and Szathmary E. 1995. The major transitions in evolution. W. H. Freeman/Spektrum, Oxford.

Mayr E. 1963. Animal species and evolution. Harvard University Press, Cambridge.

Mayr G. 2017. The Early Eocene birds of the Messel fossil site: A 48-million-year-old bird community adds a temporal perspective to the evolution of tropical avifaunas. Biol. Revs. 92: 1174–1188.

McAdam SA and Brodribb TJ. 2012. Stomatal innovation and the rise of seed plants. Ecol. Lett. 15: 1–8.

McClain CR, Balk MA, Benfield MC et al. 2015. Sizing ocean giants: Patterns of intraspecific size variation in marine megafauna. PeerJ 2: 5715.

McCracken GF, Safi K, Kunz TH et al. 2016. Airplane documents the fastest flight speeds recorded for bats. Roy. Soc. Open Sci. 3: 16–398.

McCutchen CW. 1977. The spinning rotation of ash and tulip tree samaras. Science 197: 691–692.

McGee YD, Faircloth BC, Borstein SR et al. 2016. Replicated divergence in cichlid radiations mirrors a major vertebrate innovation. Proc. Roy. Soc. B 283: 2015.1413.

McLean RB. 1974. Direct shell acquisition by hermit crabs from gastropods. Experientia 30: 206–208.

McLean RB. 1983. Gastropod shells: A dynamic resource that helps shape benthic community structure. J. Exp. Mar. Biol. Ecol. 69: 151–174.

McNeil WH. 1982. The pursuit of power: Technology, armed force, and society since A.D. 1000. University of Chicago Press, Chicago.

McShea DW. 2001. The minor transitions in hierarchical evolution and a question of a directional bias. J. Evol. Biol. 14: 502–518.

McShea DW. 2016. Three trends in the history of life: An evolutionary syndrome. Evol. Biol. 43: 531–542.

McShea DW and Brandon RN. 2010. Biology's first law: The tendency for diversity and complexity to increase in evolutionary systems. University of Chicago Press, Chicago.

Mehta RS and Wainwright PC. 2007. Raptorial jaws in the throat help moray eels swallow large prey. Nature 479: 79–82.

Melillo JM, McGuire ID, Kicklighter DW et al. 1993. Global climate change and global net primary production. Nature 363: 234–240.

Melstrom KM, Chiappe LM, and Smith ND. 2021. Exceptionally simple, rapidly replaced teeth in sauropod dinosaurs demonstrate a novel evolutionary strategy for herbivory in Late Jurassic ecosystem. BMC Evol. Ecol. 21: 202.

Meyer-Vernet N and Rospars J-P. 2017. Maximum relative speeds of living organisms: Why do bacteria perform as fast as ostriches? Physical Biol. 13: 066006.

Mills DB. 2020. The origin of phagocytosis in Earth history. Interface Focus 10: 2020.0019.

Miranda EB, de Meneses JFS, and Rheingantz ML. 2016. Reptiles as principal prey? Adaptations for durophagy and prey selection by jaguar (*Panthera onca*). J. Nat. Hist. 50: 2021–2035.

Mironenko AA. 2020. Endocerids: Suspension feeding nautiloids? Hist. Biol. 32: 281–289.

Mironenko AA. 2021. Early Palaeozoic Discodarina: A key to the appearance of cephalopod jaws. Lethala 54: 457–476.

Mitchell JG. 2002. The energetics and scaling of search strategies in bacteria. Amer. Nat. 169: 727–740.

Mitchell RL, Strullu-Derrien C, Sykes D et al. 2021. Cryptogamic ground covers as analogues for early terrestrial biospheres: Initiation and evolution of biologically mediated proto-soils. Geobiology 19: 292–306.

Mohr W, Lehnen N, Ahmerkamp S et al. 2021. Terrestrial-type nitrogen-fixing symbiosis between seagrass and a marine bacterium. Nature 600: 105–109.

Mokyr J. 1990. The lever of riches: Technological creativity and economic progress. Oxford University Press, New York.

Monod J. 1971. Chance and necessity. Alfred A. Knopf, New York.

Moral Sachetti JF del, Camacaro FIL, Vázquez JS, and Cárdenas RZ. 2011. Fuerza de mordedura y estrés mandibular en el jaguar (*Panthera onca*) durance la depredación de pecaríes (Artiodactyla: Tayassuldae) mediante la fractura de sus cráneos. Acta Zoola Mex. 27: 757–776.

Morse DH. 1975. Ecological aspects of adaptive radiation in birds. Biol. Revs. 50: 167–214.

Morton ES. 1978. Avian arboreal folivores: Why not? In Montgomery GG (ed.), The ecology of arboreal folivores. Smithsonian Press, Washington, DC: 123–130.

Motani R. 2002. Scaling effects in caudal propulsion and the speed of ichthyosaurs. Nature 415: 309–312.

Motani R and Vermeij GJ. 2021. Ecophysiological steps of marine adaptation in extant and extinct non-avian tetrapods. Biol. Revs. 96 : 1769–1798.

Muijres FT, Chang SW, van Veen WG et al. 2017. Escaping blood-fed malaria mosquitoes minimize tactile detection without compromising on take-off speed. J. Exp. Biol. 220: 3751–3762.

Muramatsu K, Yamamoto J, Abe T et al. 2013. Oceanic squid do fly. Mar. Biol. 160: 1171–1175.

Murtaza N. 2011. Pursuing self-interest or self-actualization? From capitalism to a steady-state, wisdom economy. Ecol. Econ. 70: 577–584.

Navarro-Lorbés P, Ruiz J, Díaz-Martínez I et al. 2021. Fast-running theropods tracks from the Early Cretaceous of La Rioja, Spain. Sci. Reps. 11: 23095.

Nelsen MP, Lücking R, Boyce CK et al. 2020. No support for the emergence of lichens prior to the evolution of vascular plants. Geobiology 18: 3–13.

Nelsen PE, DiMiche l e WA, Peters SE, and Boyce CK. 2016. Delayed fungal evolution did not cause the Paleozoic peak in coal production. Proc. Nat. Acad. Sci. U.S.A. 113: 2442–2447.

Neubauer S, Hublin J-J, and Gunz P. 2018. The evolution of modern human brain shape. Sci. Adv. 4: eaaz5961.

Niklas KJ. 1983. The influence of Paleozoic ovule and cupule morphologies on wind pollination. Evolution 37: 868–986.

Niklas KJ and Newman SA. 2013. The origins of multicellular organisms. Evol. Devel. 15: 41–52.

Noè F, Taylor MA, and Gómez-Pérez M. 2017. An integrated approach to understanding the long of the long neck in plesiosaurs. Acta Palaeont. Polon. 62: 137–162.

Norberg RÅ. 1973. Autorotation, self-stability, and structure of single-winged fruits (samaras) with comparative remarks on animal flight. Biol. Revs. 48: 561–596.

Nowak MA and Ohtsuki H. 2008. Prevolutionary dynamics and the origin of evolution. Proc. Nat. Acad. Sci. U.S.A. 105: 14924–14927.

Nowak MA, Tarnita CE, and Wilson EO. 2010. The evolution of eusociality. Nature 466: 1057–1062.

O'Dor R, Stewart J, Gilly W et al. 2013. Squid rocket science: How squid launch into air. Deep Sea Res. Part 11, Topical Stud. Oceanogr. 95: 113–118.

Oka S, Tomita T, and Miyamoto K. 2016. A mighty claw, pinching force of the coconut crab, the largest terrestrial crustacean. PLoS 11: e0166108.

Olivera BM. 2002. Conus venom peptides: Reflections from the biology of clades and species. Ann. Rev. Ecol. Evol. Syst. 33: 25–47.

Oliverio M and Modica MV. 2010. Relationships of the haematophagous marine snail *Colubraria* (Rachiglossa: Colubrariidae), within the neogastropod phylogenetic framework. Zool. J. Linn. Soc. 158: 779–800.

Onstein OE, Kissling WD, and Linder HP. 2022. The megafaunal gap after the non-avian dinosaur extinctions modified trait evolution and diversification of tropical palms. Proc. Roy. Soc. B 289: 2021.2633.

Ortega-Hernández J, Esteve J, and Butterfield NJ. 2013. Humble origins for a successful strategy: Complete enrolment in Early Cambrian olenellid trilobites. Biol. Lett. 9: 2013.0679.

Osborne CP and Beerling DJ. 2002. Sensitivity of tree growth to a high CO_2 environment: Consequences for interpreting the characteristics of fossil woods from ancient "greenhouse" worlds. Palaeogeogr., Palaeoclimatol., Palaeoecol. 182: 15–29.

Osborne CP and Sack L. 2012. Evolution of C_4 plants: A new hypothesis for an interaction of CO_2 and water relations mediated by plant hydraulics. Phil. Trans. Roy. Soc. B 367: 583–600.

Ossó À. 2016. *Ecogeryon elegius* n. gen. and n. sp. (Decapoda: Eubrachyura: Portunoidea), one of the oldest modern crabs from late Cenomanian of the Iberian Peninsula. Bol. Soc. Geol. Mex. 68: 231–246.

Oudman T and Piersma T. 2018. De ontsnapping van de natuur: een nieuwe kijk op kennis. AthenaeumPolak & Van Gennep, Amsterdam.

Owen DF. 1980. How plants may benefit from the animals that eat them. Oikos 35: 230–235.

Padian K. 2017. Evolution: Parallel lives. Nature 548: 156–157.

Paladino FV, O'Connor MP, and Spotila JR. 1990. Metabolism of leatherback turtles, gigantothermy, and thermoregulation of dinosaurs. Nature 344: 858–860.

Palmer AR. 2004. Symmetry breaking and the evolution of development. Science 306: 828–833.

Patek SN and Caldwell RL. 2005. Extreme impact and cavitation forces of a biological hammer: Strike forces of the peacock mantis shrimp *Odontodactylus scyllarus*. J. Exp. Biol. 208: 3655–3664.

Patek SN, Korff WL, and Caldwell RL. 2004. Deadly strike mechanism of a mantis shrimp. Nature 428: 819.

Pates S and Bicknell RDC. 2019. Elongated thoracic spines as potential predatory deterrents in olenellid trilobites of the Lower Cambrian of Nevada. Palaeogeogr., Palaeoclimatol., Palaeoecol. 516: 295–306.

Payne JL, Bachan A, Heim NA et al. 2020. The evolution of complex life and the stabilization of the Earth system. Interface Focus 10: 2019.0106.

Payne JL, Boyer AG, Brown JH et al. 2009. Two-phase increase in the maximum size of life over 3.5 billion years reflect biological innovation and environmental opportunity. Proc. Nat. Acad. Sci. U.S.A. 106: 24–27.

Payne JL, Bush AM, Heim NA et al. 2016. Ecological selectivity of the emerging mass extinction in the oceans. Science 353: 1284–1286.

Peel JS. 2015. Failed predation, commensalism and parasitism on Lower Cambrian linguliformean brachiopods. Alcheringa 39: 149–163.

Pei R, Pittman M, Goloboff P et al. 2020. Potential for powered flight neared by most close avialan relatives, but few cross its thresholds. Curr. Biol. 30: 4033–4046.

Perleman JE, Scott J, Chumley T, and Kirby JR. 2008. Predataxis behavior in *Myxococcus xanthus*. Proc. Nat. Acad. Sci. U.S.A. 105: 17127–17132.

Persons WS and Currie PJ. 2017. The functional origin of dinosaur bipedalism: Cumulative evidence from bipedally inclined reptiles and disinclined mammals. J. Theoret. Biol. 420: 1–7.

Peters RS, Krogmann L, Mayer C et al. 2017. Evolutionary history of the Hymenoptera. Curr. Biol. 27: 1013–1018.

Pfeffer SE, Wahl VL, Wittlinger M, and Wolf H. 2019. High-speed locomotion in the Saharan silver ant, *Cataglyphus bombycina*. J. Exp. Biol. 222: jeb198705.

Piketty T. 2014. Capital in the twenty-first century. Belknap Press of Harvard University, Cambridge.

Piketty T. 2020. Capital and ideology. Belknap Press of Harvard University, Cambridge.

Pinkalski C, Damgaard C, Jensen KMV et al. 2015. Non-destructive biomass estimation of *Oecophylla smaragdina* colonies: A model species for the ecological impact of ants. Insect Conserve Div. 8: 464–473.

Pinker S. 2011. The better angels of our nature: Why violence has decreased. Penguin, New York.

Pinker S. 2018. Enlightenment now: The case for reason, science, humanism, and progress. Viking, New York.

Pitman RL and Durban JW. 2012. Cooperative hunting behavior, prey selectivity and prey handling by pack ice killer whales (*Orcinus orca*), type B, in Antarctica Peninsula waters. Mar. Mammal Sci. 28: 16–36.

Planavsky NJ, Asael D, Hofmann A et al. 2014. Evidence for oxygenic photosynthesis half a billion years before the Great Oxidation Event. Nature Geosci. 7: 283–286.

Polin M, Tuval I, Drescher K et al. 2009. *Chlamydomonas* swims with two "gears" in a eukaryotic version of run-and-tumble locomotion. Science 325: 487–490.

Polis GA, Anderson WB, and Holt RD. 1997. Toward an integration of landscape and food web ecologies: The dynamics of spatially subsidized food webs. Ann. Rev. Ecol. Syst. 28: 289–316.

Ponce de León MS, Bienvenu T, Marom A et al. 2021. The primitive brain of early *Homo*. Science 372: 165–171.

Pontzer H, Brown MH, Raichlen DA et al. 2016. Metabolic acceleration and the evolution of human brain size and life history. Nature 533: 390–392.

Poore AGB, Campbell AH, Coleman RA et al. 2012. Global patterns in the impact of marine herbivores on marine primary producers. Ecol. Lett. 15: 912–922.

Popper KR. 1964. The poverty of historicism (4th edition). Harper and Row, New York.

Por FD. 1994. Animal achievement: A unifying theory of zoology. Balaban Publishers, Rehobot.

Porter SM. 2016. Tiny vampires in ancient seas: Evidence for predation via perforation in fossils from the 780–740 million-year-old Chuar Group, Grand Canyon, USA. Proc. Roy. Soc. B 283: 2016.0221.

Powell R and Mariscal C. 2015. Convergent evolution as natural experiment: The tape of life reconsidered. Interface Focus 5: 2015.0040.

Price PW. 1980. Evolutionary biology of parasites. Princeton University Press, Princeton.

Prum RO. 2017. The evolution of beauty: How Darwin's forgotten theory of mate choice shapes the animal world—and us. Doubleday, New York.

Pyenson ND and Vermeij GJ. 2016. The rise of ocean giants: Maximum body size in Cenozoic marine mammals as an indicator for productivity in the Pacific and Atlantic Oceans. Biol. Lett. 12: 2016.0186.

Raby CR, Alexis DM, Dickinson A, and Clayton MS. 2007. Planning for the future by Western scrub-jays. Nature 445: 919–921.

Raghanti PA, Edler Stephenson AR et al. 2018. A neurochemical hypothesis for the origin of hominids. Proc. Nat. Acad. Sci. U.S.A. 115: e1108–e1116.

Rahman IA, Zamora S, Falkingham PL, and Phillips JC. 2015. Cambrian cinctan echinoderms shed light on feeding in the ancestral deuterostome. Proc. Roy. Soc. B 282: 2015.1964.

Raven JA and Sánchez-Baracaldo P. 2021. *Gloeobacter* and the implications of a freshwater origin of cyanobacteria. Phycologia 60: 402–418.

Rayfield EJ, Norman DB, Horner CC et al. 2001. Cranial design and function in a large theropod dinosaur. Nature 409: 1033–1037.

Reeve HK and Hölldobler B. 2007. The emergence of a superorganism through intergroup competition. Proc. Nat. Acad. Sci. U.S.A. 104: 9736–9740.

Reich RB. 2015. Saving capitalism: For the many, not the few. Alfred A. Knopf, New York.

Retallack GJ. 2015. Silurian vegetation stature and density inferred from fossil soils and plants in Pennsylvania, USA. J. Geol. Soc. 172: 693–709.

Retallack GJ and Landing E. 2014. Affinities and architecture of Devonian trunks of *Prototaxites loganii*. Mycologia 106: 1143–1158.

Richardson AE, Chen J, Johnston R et al. 2021. Evolution of the grass leaf by promordium extension and petiole-lamina remodeling. Science 374: 1377–1381.

Ridgwell A and Zeebe RE. 2005. The role of the global carbonate cycle in the regulation and evolution of the Earth system. Earth Planet. Sci. Lett. 234: 299–315.

Rieppel O, Li C, and Fraser NC. 2008. The skeletal anatomy of the Triassic protorosaur *Dinocephalosaurus orientalis* Li, from the Middle Triassic of Guizhou Province, southern China. J. Vert. Paleont. 28: 95–110.

Roach NB, Venkadesan M, Rainbow MG, and Lieberman DE. 2013. Elastic energy storage in the shoulder and the evolution of high-speed throwing in *Homo*. Nature 498: 483–486.

Robinson GE and Barron AB. 2017. Epigenetics and the evolution of instincts. Science 356: 26–27.

Robinson JM. 1990. The burial of organic carbon as affected by the evolution of land plants. Hist. Biol. 3: 189–201.

Rodrigues TM and Machado SR. 2007. The pulvinus endodermal cells and their relation to leaf movement in legumes of the Brazilian cerrado. Plant Biol. 9: 469–477.

Rodríguez-Tovar F, Lowery CM, Bralower TJ et al. 2020. Rapid macrobenthic diversification and stabilization after the end-Cretaceous mass extinction event. Geology 48: 1048–1052.

Roman J and McCarthy JJ. 2010. The whale pump: Marine mammals enhance primary productivity in a coastal basin. PLoS One 5: e13255.

Roman J, Estes JA, Morrissette L et al. 2014. Whales as marine ecosystem engineers. Front. Ecol. Env. 12: 377–385.

Romer PM. 1986. Increasing returns and long-run growth. J. Polit. Econ. 94: 1002–1037.

Roopnarine PD. 2006. Extinction cascades and catastrophe in ancient food webs. Paleobiology 32: 1–19.

Roopnarine PD and Angielczyk YO. 2015. Community stability and selective extinction during the Permian-Triassic mass extinction. Science 350: 90–93.

Roopnarine PD, Angielczyk YO, Wang SC, and Hertog R. 2007. Trophic network models explain instability of Early Triassic terrestrial communities. Proc. Roy. Soc. B 274: 2077–2086.

Rosing MT, Bird DK, Sleep NH et al. 2006. The rise of continents—an essay on the geologic consequences of photosynthesis. Palaeogeogr., Palaeoclimatol., Palaeoecol. 132: 99–113.

Rubalcaba JG, Verberk WEP, Hendriks AJ et al. 2020. Oxygen limitation may affect the temperature and size dependence of metabolism in aquatic ectotherms. Proc. Nat. Acad. Sci. U.S.A. 117: 31963–31968.

Ruben J. 1995. The evolution of endothermy in mammals and birds: From physiology to fossils. Ann. Rev. Physiol. 57: 69–95.

Rubenstein DI and Koehl MAR. 1977. The mechanisms of filter feeding: Some theoretical considerations. Amer. Nat. 111: 981–994.

Ruxton GD and Wilkinson DM. 2011. The energetics of low browsing in sauropods. Biol. Lett. 7: 779–781.

Rybczynski N and Reisz RR. 2001. Earliest evidence for efficient oral processing in a terrestrial herbivore. Nature 411: 684–686.

Sachs G, Traugott J, Nesterova AP, and Bonadonna F. 2013. Experimental verification of dynamic soaring in albatrosses. J. Ex. Biol. 216: 4222–4232.

Salas-Gismondi R, Flynn JJ, Baby P et al. 2015. A Miocene hyperdiverse crocodylian community reveals peculiar trophic dynamics in proto-Amazonian mega-wetlands. Proc. Roy. Soc. B 282: 2014.2492.

Sander PM, Griebeler DM, Klein N, et al. 2021. Early giant reveals faster evolution of large body size in ichthyosaurs than in cetaceans. Science 374: eabf5787.

Sato K. 2014. Body temperature stability achieved by the large body mass of sea turtles. J. Exp. Biol. 217: 36-7-3614.

Satyanath S, Voigtländer N, and Voth H-J. 2017. Bowling for Fascism: Social capital and the rise of the Nazi Party. J. Polit. Econ. 125: 478–526.

Saulsbury J. 2020. Crinoid respiration and the distribution of energetic strategies among marine invertebrates. Biol. J. Linn. Soc. 129: 244–258.

Savoca MS, Czapanskiy MF, Kahane-Rapport SR et al. 2021. Baleen whale prey consumption based on high-resolution foraging measurements. Nature 599: 85–90.

Savoca MS and Nevitt GA. 2014. Evidence that dimethylsulfide facilitates a tritrophic mutualism between marine primary producers and top predators. Proc. Nat. Acad. Sci. U.S.A. 111: 4157–4161.

Schaeffer B and Rosen DE. 1961. Major adaptive levels in the evolution of the actinopterygian feeding mechanism. Amer. Zool. 1: 187–204.

Schaller GB. 1972. The Serengeti lion: A study of predator-prey relations. University of Chicago Press, Chicago.

Scheffer M, van Bavel B, van de Leemput IA, and van Nes EH. 2017. Inequality in nature and society. Proc. Nat. Acad. Sci. U.S.A. 114: 13154–13157.

Schmidt M and Dejean A. 2018. A dolichoderine ant that constructs traps to ambush prey collectively: Convergent evolution with a myrmecine genus. Biol. J. Linn. Soc. 124: 41–46.

Schrorger AW. 1941. The bronzed grackle's method of opening acorns. Wilson Bull. 53: 138–140.

Schulz JR, Jan I, Sanva G, and Azizi E. 2019. The high speed radular prey strike of a fish-hunting cone snail. Curr. Biol. 29: R788–R789.

Schulz KG, Zondervan I, Gerringa LJA et al. 2004. Effect of trace metal availability on coccolithophorid calcification. Nature 430: 373–376.

Schwarz D, Fedler MT, Lukas P, and Kupfer A. 2021. Form and function of the feeding apparatus of sirenid salamanders (Caudata: Sirenidae): Three-dimensional chewing and herbivory? Zool. Anz. 295: 99–116.

Seabright P. 2004. The company of strangers: A natural history of economic life. Princeton University Press, Princeton.

Sellers WI and Manning PL. 2007. Estimating dinosaur maximum running speeds using evolutionary robotics. Proc. Roy. Soc. B s74: 2711–2716.

Sereno PC, Larsson HCE, Sidor CA, and Gado B. 2001. The giant crocodyliform *Sarcosuchus* from the Cretaceous of Africa. Science 294: 1516–1519.

Seymour RS and Schultze-Motel P. 1996. Thermoregulating lotus flowers. Nature 383: 305.

Seymour RS, White CR, and Gibernau M. 2003. Heat reward for insect pollinators. Nature 426: 243–244.

Sharoni S and Halevy I. 2022. Geological controls on phytoplankton elemental composition. Proc. Nat. Acad. Sci. U.S.A. 119: e211323118.

Shimada K. 2021. The size of the megatooth shark, *Otodus megalodon* (Lamniformes: Otodontidae), revisited. Hist. Biol. 33: 904–911.

Shimek RL and Kohn AJ. 1981. Functional morphology and evolution of the toxoglossan radula. Malacologia 20: 423–438.

Shurin JB, Gruner DS, and Hillebrand H. 2006. All wet or dried up? Real differences between aquatic and terrestrial food webs. Proc. Roy. Soc. B 273: 1–9.

Sicangon CK, Lavdekar S, Subhash G, and Putz FE. 2022. Active space garnering by leaves of a rosette plant. Curr. Biol. 32: R352–353.

Signor PW III and Brett CE. 1984. The mid-Paleozoic precursor to the Mesozoic Marine Revolution. Paleobiology 10: 229–245.

Signor PW and Vermeij GJ. 1994. The plankton and the benthos: Origins and early history of an evolving relationship. Paleobiology 20: 297–319.

Sigwart JD, Vermeij GJ, and Hoyer P. 2019. Why do chitons curl into a ball? Biol. Lett. 15: 2019.0429.

Sillett SC, Van Pelt R, Koch GW et al. 2010. Increasing wood production through old age in tall trees. Forest Ecol. and Management 259: 976–994.

Simon JL. 1977. The economics of population growth. Princeton University Press, Princeton.

Sinninghe Damsté JS, Muyzer G, Abbas B et al. 2004. The rise of the rhizosolenid diatoms. Science 304: 584–587.

Skotheim J and Mahadevan L. 2005. Physical limits and design principles for plant and fungal movements. Science 308: 1308–1310.

Slater GJ, Goldbogen J, and Pyenson ND. 2017. Independent evolution of baleen whale gigantism linked to Plio-Pleistocene ocean dynamics. Proc. Roy. Soc. B 284: 2017.0546.

Slater GJ and Van Valkenburgh B. 2008. Long in the tooth: Evolution of sabertooth cat cranial shape. Paleobiology 34: 403–419.

Smetacek V. 2021. A whale of an appetite. Nature 599: 33–34.

Smith A. 1776. An inquiry into the nature and causes of the wealth of nations (1971 edition, with introduction by Cannon E and notes by Lerner M). Random House, New York.

Smith AR. 1973. Stratification of temperate and tropical forests. Amer. Nat. 107: 671–683.

Smith CPA, Laville T, and Fara E. 2021. Exceptional fossil assemblages confirm the existence of complex Early Triassic ecosystems during the early Spathian. Sci. Reps 11: 19657.

Smith DC and Bernays EA. 1991. Why do so few animals form endosymbiotic associations with photosynthetic microbes? Phil. Trans. Roy. Soc. London B 333: 225–230.

Smith FA, Payne JL, Heim NA et al. 2016. Body size evolution across the Geozoic. Ann. Rev. Earth Planet. Sci. 44: 523–553.

Smith WL, Stern JH, Girard MG, and Davis IQ. 2016. Evolution of venomous and cartilaginous and ray-finned fishes. Integr. Comp. Biol. 56: 950–961.

Snively E, Fahlke JM, and Welsh RC. 2015. Bone-breaking bite force of *Basilosaurus isis* (Mammalia, Cetacea) from the Eocene of Egypt estimated by finite element analysis. PLoS One 10: e0118380.

Snively E, Russell AP, Powell GL et al. 2014. The role of the neck in the feeding behaviour of the Tyrannosauridae: Inference based on kinematics and muscle function of extant avians. J. Zool. 292: 290–303.

Snooks CD. 1998. The laws of history. Routledge, London.

Somjee U, Woods HA, Duell M, and Miller CW. 2018. The hidden cost of sexually selected traits: The metabolic expense of maintaining a sexually selected weapon. Proc. Roy. Soc. B 285: 2018.1685.

Sorkin B. 2008. A biomechanical constraint on body mass in terrestrial mammalian predators. Lethaia 41: 333–347.

Spenkuch JL and Tillmann P. 2018. Elite influence? Religion and the electoral success of the Nazis. Amer. J. Polit. Sci. 62:19–36.

Springer AM, Estes JA, Van Vliet GB et al. 2003. Sequential megafaunal collapse in the North Pacific Ocean: An ongoing legacy of industrial whaling? Proc. Nat. Acad. Sci. U.S.A. 100: 12223–12228.

Stanley SM. 1968. Post-Paleozoic adaptive radiation of infaunal bivalve molluscs—a consequence of mantle fusion and siphon formation. J. Paleont. 42: 214–229.

Stanley SM. 1973. An explanation for Cope's Rule. Evolution 28: 1–26.

Stanley SM. 1982. Gastropod torsion: Predation and the opercular imperative. Neues Jb. Geol. Paläont. Abh. 164: 95–107.

Stanley SM. 1990. Delayed recovery and the pacing of major extinctions. Paleobiology 16: 401–414.

Steele-Petrovic HM. 1979. The physiological differences between articulate brachiopods and filter-feeding bivalves as a factor in the evolution of marine level-bottom communities. Palaeontology 22: 101–134.

Stevenson RA, Evangelista D, and Looy CV. 2015. When conifers took flight: A biomechanical evaluation of an imperfect evolutionary takeoff. Paleobiology 41: 205–225.

Stewart HL and Carpenter RC. 2003. The effects of morphology and water flow on photosynthesis of marine macroalgae. Ecology 84: 2999–3012.

Stiefel KM. 2021. Evolutionary trends in large pelagic filter-feeders. Hist. Biol. 33: 1477–1488.

Stocker R and Durham WM. 2009. Tumbling and stealth. Science 325: 400–402.

Stockey RG, Pohl A, Ridgwell A et al. 2021. Decreasing Phanerozoic extinction intensity as a consequence of Earth surface oxygenation and the metazoan ecophysiology. Proc. Nat. Acad. Sci. U.S.A. 118: e2101900118.

Strassmann JE and Queller DC. 2007. Insect societies as divided organisms: The complexities of purpose and cross-purpose. Proc. Nat. Acad. Sci. U.S.A. 104: 8619–8626.

Strathmann RR. 1990. Why life histories evolve differently in the sea. Amer. Zool. 30: 1979207.

Strevens M. 2020. The knowledge machine: How irrationality created modern science. New York: Liveright.

Strother PK and Foster C. 2021. A fossil record of land plant origins from charophyte algae. Science 373: 792–796.

Strullu-Derrlen C, Selosse M-A, Kenrick P, and Martin F. 2018. The origin and evolution of mycorrhizal symbioses: From palaeomycology to phylogenomics. New Phytologist 220: 1012–1030.

Suarez PA and Leys SP. 2022. The sponge pump as a morphological character in the fossil record. Paleobiology 48: 446–461.

Sulej T and Niedźwiedcki J. 2019. An elephant-sized Late Triassic synapsid. Science 363: 78–80.

Sunagar K and Moran Y. 2015. The rise and fall of an evolutionary innovation: Contrasting strategies of venom evolution in ancient and young animals. PLoS Genetics 11: e1005596.

Syvitski J, Waters CN, Day J et al. 2020. Extraordinary human energy consumption and resultant geological impacts beginning around 1950 CE initiated the proposed Anthropocene epoch. Commun. Earth and Evo. 1: 32.

Tanaka K, Anvarov OUO, Zelenitsky DK et al. 2021. A new carcharodontosaurian theropod dinosaur occupies apex predator niche in the early late Cretaceous of Uzbekistan. Roy. Soc. Open Sci. 8: 210923.

Tanner W and Beevers H. 2001. Transpiration, a prerequisite for long-distance transport of minerals in plants? Proc. Nat. Acad. Sci. U.S.A. 98: 9443–9447.

Tarhan LG, Droser ML, Planavsky NJ, and Johnston DT. 2015. Protracted development of bioturbation through the Early Palaeozoic era. Nature Geosci. 8: 865–869.

Tarhan LG, Zhao M, and Planavsky NJ. 2021. Bioturbation feedbacks on the phosphorus cycle. Earth Planet. Sci. Lett. 566: 116961.

Thayer CW. 1979. Biological bulldozers and the evolution of marine benthic communities. Science 203: 458–461.

Thayer CW. 1983. Sediment-mediated biological disturbance and the evolution of marine benthos. In Tevesz MJS and McCall PL (eds.), Biotic interactions in Recent and fossil benthic communities. Plenum, New York: 479–625.

Thiery G and Sha JC. 2020. Low occurrence of molar use in black-tufted capuchin monkeys: Should adaptation to seed ingestion be inferred from molars in primates? Palaeogeogr., Palaeoclimatol., Palaeoecol. 555: 109853.

Thompson DW. 1942. On growth and form. Cambridge University Press, London.

Tiffney BH. 1984. Seed size, dispersal syndromes, and the rise of the angiosperms: Evidence and hypotheses. Ann. Missouri Bot. Garden 71: 551–576.

Tilman D, Reich PB, and Isbell F. 2012. Biodiversity impacts ecosystem productivity as much as resources, disturbance, or herbivory. Proc. Nat. Acad. Sci. U.S.A. 109: 10394–10397.

Timerman D and Barrett SCH. 2021. The biomechanics of pollen release: New perspectives on the evolution of wind pollination in angiosperms. Biol. Revs. 96: 2156–2161.

Timm-Davis LL, Davis RW, and Marshall CD. 2017. Durophagous biting in sea otters (*Enhydra lutris*) differs kinematically from raptorial biting of other marine mammals. J. Exp. Biol. 220: 4703–4710.

Titus L, Knoll K, Sertich JJW et al. 2021. Geology and taphonomy of a unique tyrannosaurid bonebed from the Upper Campanian Kaiparowits Formation of southern Utah: Implications for tyrannosaurid gregariousness. PeerJ 9: e11013.

Tomback DF and Linhart YB. 1990. The evolution of bird-dispersed seeds. Evol. Ecol. 4: 185–219.

Tostevin R, Wood RA, Shields GA et al. 2016. Low-oxygen waters limited habitable space for early animals. Nature Comm. 7: 12818.

Trueblood LA and Seibel BA. 2014. Slow swimming, fast strikes: Effects of feeding behavior on scaling of anaerobic metabolism in epipelagic squid. J. Exp. Biol. 217: 2710–2716.

Tsai C-H and Kohno N. 2016. Multiple origins of gigantism in stem baleen whales. The Science of Nature 103: 89.

Tseng ZJ, Antón M, and Salesa MJ. 2011. The evolution of the bone-cracking model in carnivores: cranial functional morphology of the Plio-Pleistocene cursorial hyaenid *Chasmaporthetes lunensis* (Mammalia: Carnivora). Paleobiology 37: 140–156.

Tseng ZJ, Grohé C, and Flynn JJ. 2016. A unique feeding strategy of the extinct marine mammal *Kolponomos*: Convergence on sabretooths and sea otters. Proc. Roy. Soc. B 283: 2016.0044.

Tseng ZJ and Wang X. 2010. Cranial functional morphology of fossil dogs and adaptation for durophagy in *Borophagus* and *Epicyon* (Carnivora, Mammalia), J. Morphol. 271: 1386–1398.

Turner ECO. 2021. Possible poriferan body fossils in early Neoproterozoic microbial reefs. Nature 596: 87–91.

Turner JS. 2007. The tinkerer's accomplice: How design emerges from life itself. Harvard University Press, Cambridge.

Ulyshen MD. 2016. Wood decomposition as influenced by invertebrates. Biol. Revs. 91: 70–85.

Valenciano A, Baskin JA, Abella J et al. 2016. *Megalictis*, the bone-crushing giant mustelid (Carnivora, Mustelidae, Oligobuninae) from the Early Miocene of North America. PLoS One 11: e0152430.

Vanderklift MA, Pillans RD, Hutton M et al. 2021. High rates of herbivory in remote northwest Australian seagrass meadows by rabbitfish and green turtles. J. Exp. Mar. Biol. Ecol. 665: 63–73.

Vander Wall SB. 2001. The evolutionary ecology of nut dispersal. Bot. Rev. 67: 75–117.

Vander Wall SB and Jenkins SH. 2003. Reciprocal pilferage and the evolution of food-hoarding behavior. Behav. Ecol. 14: 656–667.

Van Roy P, Daley AC, and Briggs DEG. 2015. Anomalocaridid trunk homology revealed by a giant filter-feeder with paired flaps. Nature 523: 77–80.

Van Valen L. 1976. Energy and evolution. Evol. Theory 1: 179–229.

Van Valkenburgh B. 1985. Locomotor diversity within past and present guilds of large predatory mammals. Paleobiology 11: 407–428.

Van Valkenburgh B. 1991. Iterative evolution of hypercarnivory in canids (Mammalia: Carnivora): Evolutionary interaction among sympatric predators. Paleobiology 17: 340–362.

Van Valkenburgh B. 1999. Major patterns in the history of carnivorous mammals. Ann. Rev. Earth Planet. Sci. 27: 463–493.

Van Valkenburgh B, Grady F, and Kurtén B. 1990. The Plio-Pleistocene cheetah-like cat *Miracinonyx inexpectatus* of North America. Vert. Paleont. 10: 434–454.

Van Valkenburgh B and Hertel F. 1993. Tough times at La Brea: Tooth breakage in large carnivores of the Late Pleistocene. Science 261: 456–459.

Van Valkenburgh B, Maddox T, Funston PJ et al. 2009. Sociality in Rancho La Brea *Smilodon*: Arguments favour "evidence" over "coincidence." Biol. Lett. 5: 563–564.

Verberk WCEP, Bilton DT, Calosi P, and Spicer JI. 2011. Oxygen supply in aquatic ectotherms: Partial pressure and solubility together explain biodiversity and size patterns. Ecology 92: 1565–1572.

Vermeij GJ. 1975. Evolution and distribution of left-handed and planispiral coiling in snails. Nature 254: 419–420.

Vermeij GJ. 1977. The Mesozoic marine revolution: Evidence from snails, predators and grazers. Paleobiology 3: 245–258.

Vermeij GJ. 1982a. Gastropod shell form, repair, and breakage in relation to predation by the crab Calappa. Malacologia 23: 1–12.

Vermeij GJ. 1982b. Unsuccessful predation and evolution. Amer. Nat. 120: 701–720.

Vermeij GJ. 1987. Evolution and escalation: An ecological history of life. Princeton University Press, Princeton.

Vermeij GJ. 1991. When biotas meet: Understanding biotic interchange. Science 253: 1099–1104.

Vermeij GJ. 1993. A natural history of shells. Princeton University Press, Princeton.

Vermeij GJ. 1995. Economics, volcanoes, and Phanerozoic revolutions. Paleobiology 21: 125–152.

Vermeij GJ. 1999. Inequality and the directionality of history. Amer. Nat. 153: 243–253.

Vermeij GJ. 2002. Characters in context: Molluscan shells and the forces that mold them. Paleobiology 28: 41–54.

Vermeij GJ. 2003. Temperature, tectonics, and evolution. In Rothschild LJ and Lister A (eds.), Evolution on planet Earth: The impact of the physical environment. Academic Press, Amsterdam: 209–232.

Vermeij GJ. 2004a. Ecological avalanches and the two kinds of extinction. Evol. Ecol. Res. 6: 315–337.

Vermeij GJ. 2004b. Nature: An economic history. Princeton University Press, Princeton.

Vermeij GJ. 2005. Invasion as expectation: A historical fact of life. In Sax DF, Stachowicz JJ, and Gaines SD (eds.), Species invasions: Insights into ecology, evolution, and biogeography. Sinauer, Sunderland: 315–339.

Vermeij GJ. 2006. Historical contingency and the purported uniqueness of evolutionary innovations. Proc. Nat. Acad. Sci. U.S.A. 103: 1804–1809.

Vermeij GJ. 2008a. Escalation and its role in Jurassic biotic history. Palaeogeogr., Paleoclimatol., Palaeoecol. 263: 3–8.

Vermeij GJ. 2008b. Security, unpredictability, and evolution: Policy and the history of life. In Sagarin RD and Taylor T (eds.), National security: A Darwinian approach to a dangerous world. University of California Press, Berkeley: 25–41.

Vermeij GJ. 2009. Comparative economics: Evolution and the modern economy. J. Bioecon. 105–134.

Vermeij GJ. 2010. The evolutionary world: How adaptation explains everything from seashells to civilization. Thomas Dunne Books, New York.

Vermeij GJ. 2011a. A historical conspiracy: Competition, opportunity, and the emergence of direction in history. Clidynamics 2: 187–207.

Vermeij GJ. 2011b. Shifting sources of productivity in the coastal marine tropics during the Cenozoic era. Proc. Roy. Soc. B 278: 2362–2368.

Vermeij GJ. 2012a. The evolution of gigantism on temperate seashores. Biol. J. Linn. Soc. 106: 776–793.

Vermeij GJ. 2012b. The limits of adaptation: Humans and the predator-prey arms race. Evolution 66: 2007–2014.

Vermeij GJ. 2013a. On escalation. Ann. Rev. Earth Planet. Sci. 41: 1–19.

Vermeij GJ. 2013b. The evolution of molluscan photosymbioses: A critical appraisal. Biol. J. Linn. Soc. 109: 497–511.

Vermeij GJ. 2015a. Forbidden phenotypes and the limits of evolution. Interface Focus 5: 2015.0028.

Vermeij GJ. 2015b. Gastropod skeletal defences: Land, freshwater, and sea compared. Vita Malacologica 13: 1–25.

Vermeij GJ. 2015c. Plants that lead: Do some surface features direct enemy traffic on leaves and stems? Biol. J. Linn. Soc. 116: 288–294.

Vermeij GJ. 2016a. Gigantism and its implications for the history of life. PLoS One 11: e0146092.

Vermeij GJ. 2016b. Plant defences on land and in water: Why are they so different? Ann. Bot. 117: 1099–1109.

Vermeij GJ. 2017a. How the land became the locus of major evolutionary innovations. Curr. Biol. 27: 3178–3182.

Vermeij GJ. 2017b. Life in the arena: Infaunal gastropods and the late Phanerozoic expansion of marine ecosystems into sand. Palaeontology 60: 649–661.

Vermeij GJ. 2018a. Building a healthy economy: Learning from nature. In Mink O and Oosterhuis W (eds.), Economia: methods for reclaiming economy. Bal tan Laboratories, Eindhoven: 53–76 and 111–119.

Vermeij GJ. 2018b. Comparative biogeography: Innovations and the rise to dominance of the North Pacific biota. Proc. Roy. Soc. B 285: 2018.2027.

Vermeij GJ. 2019a. The efficiency paradox: How wasteful competitors forged thrifty ecosystems. Proc. Nat. Acad. Sci. U.S.A. 116: 17619–17623.

Vermeij GJ. 2019b. Power, competition, and the nature of history. Paleobiology 45: 517–530.

Vermeij GJ. 2020. Getting out of arms way: Star wars and snails on the seashore. Biol. Bull. 239: 209–217.

Vermeij GJ. 2022a. Are saltmarshes younger than mangrove swamps? Ecol. Evol. 12: e8481.

Vermeij GJ. 2022b. The balanced life: Evolution of ventral shell weighting in gastropods. Zool. J. Linn. Soc. 194: 256–275.

Vermeij GJ. 2023. The origin and evolution of human uniqueness. In H Desmon and G Ramsey (eds.), Human success: Evolutionary origins and ethical implications. Oxford University Press, Oxford.

Vermeij GJ and Grosberg RK. 2010. The great divergence: When did diversity on land exceed that in the sea? Integr. Comp. Biol. 675–682.

Vermeij GJ and Leigh EG Jr. 2011. Natural and human economies compared. Ecosphere 2: 39.

Vermeij GJ and Lindberg DR. 2000. Delayed herbivory and the assembly of marine benthic ecosystems. Paleobiology 26: 419–430.

Vermeij GJ and Motani R. 2018. Land to sea transitions in vertebrates: The dynamics of colonization. Paleobiology 44: 237–258.

Vermeij GJ and Roopnarine PD. 2013. Reining in the Red Queen: The dynamics of adaptation and extinction reexamined. Paleobiology 39: 560–575.

Vermeij GJ, Schindel DE, and Zipser E. 1981. Predation through geological time: Evidence from gastropod shell repair. Science 214: 1024–1026.

Vermeij GJ and Zipser E. 2015. The diet of *Diodon hystrix* (Teleostei: Tetraodontiformes): Shell-crushing on Guam's reefs. Bishop Mus. Bull. Zool. 9: 169–175.

Vinther J, Stein K, Longrich NR, and Harper DAT. 2014. A suspension-feeding anomalocarid from the Early Cambrian. Nature 504: 496–499.

Virot E, Ma G, Clanet C, and Jung S. 2017. Physics of chewing in terrestrial mammals. Sci. Reps. 7: 43967.

Vizcaíno SF and Fariña RA. 1999. On the flight capabilities and distribution of the giant Miocene bird *Argentavis magnificens*. Lethaia 32: 271–278.

Wagner GP. 2014. Homology, genes, and evolutionary innovation. Princeton University Press, Princeton.

Walde M, Allan E, Cappelli SL et al. 2021. Both diversity and functional composition affect productivity and water use efficiency in experimental temperate grasslands. J. Ecol. 109: 3877–3891.

Wallace MW, Hood AS, Woon EMS et al. 2014. Enigmatic chambered structures in Cryogenian reefs: The oldest sponge-grade organisms? Precambrian Res. 255: 109–123.

Wallace MW, Hood AS, Shuster A et al. 2017. Oxygenation history of the Neoproterozoic to early Phanerozoic and the rise of land plants. Earth and Planet. Sci. Lett. 466: 12–16.

Ward LM, Rasmussen B, and Fischer WW. 2019. Primary productivity was limited by electron donors prior to the advent of oxygenic photosynthesis. J. Geophys. Res.: Biogeosciences 124: 211–226.

Wardle CS, Videler JJ, Arimoto T et al. 1989. The muscle twitch and the maximum swimming speed of giant bluefin tuna, *Thunnus thynnus* L. J. Fish Biol. 35: 129–137.

Wegner NC, Snodgrass OE, Dewar H, and Hyde JR. 2015. Whole-body endothermy in a mesopelagic fish, the opah, *Lampris guttatus*. Science 348: 786–789.

Weimerskirch H, Bishop C, Jeanniard-du Dot T et al. 2016. Frigate birds track atmospheric conditions over months-long transoceanic flights. Science 353: 774–778.

Weishampel DB and Jianu C-M. 2000. Plant-eaters and ghost lineages: Dinosaurian herbivory revisited. In Sues H-D (ed.), Evolution of herbivory in terrestrial vertebrates: perspectives from the fossil record. Cambridge University Press, Cambridge: 123–143.

Wellington GM. 1980. Reversal of digestive interactions between Pacific reef corals: Mediation by sweeper tentacles. Oecologia 47: 340–343.

Wells MJ and O'Dor RK. 1991. Jet propulsion and the evolution of the cephalopods. Bull. Mar. Sci. 49: 419–432.

Welsh JQ and Bellwood DR. 2014. Herbivorous fishes, ecosystem function and mobile links on coral reefs. Coral Reefs 33: 303–311.

West GB, Brown JH, and Enquist BJ. 1997. A general model for the origin of allometric scaling laws in biology. Science 276: 122–126.

West SA, Fisher RM, Gardner A, and Kiers ET. 2015. Major evolutionary transitions in individuality. Proc. Nat. Acad. Sci. U.S.A. 112: 10112–10119.

West-Eberhard MJ. 1983. Sexual selection, social competition, and speciation. Quart. Rev. Biol. 58: 155–183.

Whalen CD and Briggs DEG. 2018. The Palaeozoic colonization of the water column and the rise of global nekton. Proc. Roy. Soc. B 285: 2018.0883.

Whalen MA, Whippo RDB, York PH et al. 2020. Climate drives the geography of marine consumption by changing predator communities. Proc. Nat. Acad. Sci. U.S.A. 117: 28160–28166.

Williams GC. 1966. Adaptation and natural selection: A critique of some current evolutionary thought. Princeton University Press, Princeton.

Williams HJ, Shepard ELC, Holton ID et al. 2020. Physical limits of flight performance in the heaviest soaring bird. Proc. Nat. Acad. Sci. U.S.A. 117: 17884–17890.

Willson MF. 1972. Seed size preference in finches. Wilson Bull. 84: 449–455.

Wilson AM, Lowe JC, Roskilly K et al. 2013. Locomotion dynamics of hunting in wild cheetahs. Nature 498: 185–189.

Wilson DS and Gowdy JM. 2015. Human ultrasociality and the invisible hand: Foundational developments in evolutionary science alter a foundational concept in economics. J. Bioecon. 17: 37–52.

Wilson EO. 2012. The social conquest of Earth. Liveright, New York.

Wilson JP, Montañez IP, White JD et al. 2017. Dynamic Carboniferous tropical forests: New views of plant function and potential for physiological forcing of climate. New Phytologist 215: 1333–1353.

Wilson JP, White JD, Montañez IP et al. 2020. Carboniferous plant physiology breaks the mold. New Phytologist 227: 667–679.

Witton MP and Habib MB. 2010. On the size and flight diversity of giant pterosaurs, the use of birds as pterosaur analogues and comments on pterosaur flightlessness. PLoS One 5: e13982.

Wood R, Ivantsov AY, and Zhuravlev AY. 2017. First macrobiota biomineralization was environmentally triggered. Proc. Roy. Soc. B 284: 2017.0059.

Wrangham R. 2017. Control of fire in the Paleolithic: Evaluating the cooking hypothesis. Curr. Anthropol. 58: S303–S313.

Wroe S, Huber DR, Lowry M et al. 2008. Three-dimensional computer analysis of white shark jaw mechanics: How hard can a great white bite? J. Zool. 277: 336–342.

Xiao S. 2014. The making of Ediacaran giants. Curr. Biol. 24: R120–R122.

Xiao S and Dong L. 2006. On the morphological and ecological history of Proterozoic macroalgae. In Xiao S and Kaufman AJ (eds.), Proterozoic geobiology and paleobiology. Springer, Dordrecht: 57–90.

Xu G-H, Zhao J-J, Gao K-Q, and Wu F-X. 2015. A new stem-neopterygian fish from the Middle Triassic of China shows the earliest over-water gliding strategy of the vertebrates. Proc. Roy. Soc. B 280: 2015.2261.

Xu Q, Jin J, and Labandeira CCV. 2018. Williamson Drive: Herbivory from a North-central Texas flora of latest Pennsylvanian age shows discrete component community structure, expansion of piercing and sucking, and plant counterdefenses. Rev. Palaeobot. Palynol. 251: 28–72.

Yamaoka K. 1978. Pharyngeal jaw structure in labrid fish. Publ. Seto Mar. Biol. Lab. 14: 409–416.

Yamazaki K. 2011. Gone with the wind: Trembling leaves may deter herbivory. Biol. J. Linn. Soc. 104: 738–747.

Yanoviak SP, Dudley R, and Kaspari M. 2005. Directed aerial descent in canopy ants. Nature 433: 624–626.

Yanoviak SP, Kaspari M, and Dudley R. 2009. Gliding hexapods and the origin of insect aerial behaviour. Biol. Lett. 5: 510–512.

Yin Z, Zhu M, Davidson EH et al. 2015. Sponge grade body fossil with cellular resolution dating 60 Myr before the Cambrian. Proc. Nat. Acad. Sci. U.S.A. 112: e1453–e1460.

Yip EC, Powers KS, and Avilés L. 2008. Cooperative capture of large prey solves scaling challenge faced by spider societies. Proc. Nat. Acad. Sci. U.S.A. 105: 11818–11822.

Zahavi A. 1975. Mate selection—a selection for a handicap. J. Theoret. Biol. 53: 205–214.

Zalasiewicz J, Williams M, Waters CN et al. 2017. Scale and diversity of the physical technosphere: A geological perspective. The Anthropocene 4: 9–22.

Zhang ZF, Holmer LE, Robson SP et al. 2011. First record of repaired durophagous shell damages in Early Cambrian lingulate brachiopods with preserved pedicles. Palaeogeogr., Palaeoclimatol., Palaeoecol. 302: 206–2127.

Zhou B-F, Yuan S, Crowl AA et al. 2022. Phylogenomic analyses highlight innovation and introgression in the continental radiation of Fagaceae across the Northern Hemisphere. Nature Comm. 13: 1320.

Zhou H, Hu Y, Ebrahlmi A et al. 2021. Whole genome based insights into the phylogeny and evolution of the Juglandaceae. BMC Evol. Ecol. 21: 190.

Zhu MY, Vannier T, van Iten E, and Zhao YL. 2004. Direct evidence for predation on trilobites in the Cambrian. Proc. Roy. Soc. London B 271: S277–S280.

Zhuravlev AY and Wood RA. 2008. Eve of biomineralization: Controls on skeletal mineralogy. Geology 36: 923–926.

Ziebis W, Forster S, Huettel M, and Jørgensen BB. 1996. Complex burrows of the mud shrimp *Callianassa truncata* and their geochemical impact in the sea bed. Nature 382: 619–622.

Zimmermann MH. 1983. Xylem structure and the ascent of sap. Springer, Berlin.

Zona S and Christenhusz MJM. 2015. Litter-trapping plants: Filter-feeders of the plant kingdom. Bot. J. Linn. Soc. 179: 553–586.

Zverkov NG and Pervushov EM. 2020. A gigantic pliosaurid from the Cenomanian (Upper Cretaceous) of the Volga region, Russia. Cretaceous Res. 120: 104410.

INDEX